Efficient Computation of Argumentation Semantics

Efficient Computation of Argumentation Semantics

Beishui Liao

ZHEJIANG UNIVERSITY PRESS
浙江大学出版社

AMSTERDAM • BOSTON • HEIDELBERG • LONDON • NEW YORK
OXFORD • PARIS • SAN DIEGO • SAN FRANCISCO
SINGAPORE • SYDNEY • TOKYO

Academic Press is an imprint of Elsevier

Academic Press is an imprint of Elsevier
The Boulevard, Langford Lane, Kidlington, Oxford OX5 1GB, UK
225 Wyman Street, Waltham, MA 02451, USA

First edition 2014

Notice
No responsibility is assumed by the publisher for any injury and/or damage to persons
or property as a matter of products liability, negligence or otherwise, or from any use or
operation of any methods, products, instructions or ideas contained in the material herein.
Because of rapid advances in the medical sciences, in particular, independent verification
of diagnoses and drug dosages should be made.

British Library Cataloguing in Publication Data
A catalogue record for this book is available from the British Library

Library of Congress Cataloging-in-Publication Data
A catalog record for this book is available from the Library of Congress

ISBN–13: 978-0-12-410406-8

For information on all Academic Press publications
visit our web site at store.elsevier.com

Printed and bound by CPI Group (UK) Ltd, Croydon, CR0 4YY
Transferred to digital print 2013

Working together
to grow libraries in
developing countries

www.elsevier.com • www.bookaid.org

Contents

Preface

In artificial intelligence, argumentation is a mechanism for reasoning with incomplete and conflicting information, by means of constructing, comparing and evaluating arguments. In existing literature, many aspects of argumentation have been widely studied, including semantics, proof theories and algorithms, concrete argumentation systems, and various applications. One of the most important problems in this area is that under most of semantics, many natural questions regarding argument acceptability are computationally intractable.

To cope with this problem, some efforts have been made, such as identifying tractable classes of argumentation frameworks, and developing efficient algorithms. However, for a generic argumentation framework, how to efficiently compute its semantics and the changing of semantics in dynamic settings is still an open problem.

Inspired by the efficiency of local computation, in recent years, we have conducted a series of research on computing the semantics of a dynamic argumentation system, a static argumentation system, and a partial argumentation system. The corresponding results have been published in some journals such as *Artificial Intelligence, Journal of Logic and Computation* and *Annals of Mathematics and Artificial Intelligence*, etc. Based on the results we have obtained, the aim of this book is to systematically formulate the approaches and algorithms for efficient computation of argumentation semantics.

Since the computation of argumentation semantics is fundamental to the abstract argumentation theory, which is in turn the basis of concrete argument systems, I believe that the contents of this book could be useful for readers who are in the areas of knowledge representation and reasoning, as well as some related application areas.

I thank Robert C Koons and Vladimir Lifschitz for their valuable discussions and comments, when I visited the University of Texas at Austin during July 2009 to July 2011. I thank Leon van der Torre, Richard Booth, Martin Caminada, and Tjitze Rienstra for the valuable discussions with them when I visited the University of Luxembourg in April 2012. And, I thank Pietro Baroni and Massimiliano Giacomin for their insightful comments on a technical error in our previous works about the decomposability of ideal semantics, when I attended the Dagtuhl Seminar on Belief Change and Argumentation in Multi-Agent Scenarios in June 2013.

I acknowledge support from the National Social Science Foundation Major Project of China under grant No.11&ZD088, the National Grand Fundamental Research 973 Program of China under grant No.2012CB316400 and the National Natural Science Foundation of China under grant No.61175058 and No.60773177.

Beishui Liao, Hangzhou, China

July 2013

Introduction

Chapter Outline

1.1 Background

In the area of artificial intelligence, one of the most important topics is to handle the uncertainty, incompleteness, and inconsistency of information. There are a lot of applications related to this topic, such as belief revision, decision-making, inter-agent communication, medical reasoning, legal reasoning, semantic web reasoning, trust computing and normative systems, etc.

When an agent is situated in an open and dynamic environment, since the world can be other than what it appears and the world changes [1], what the agent perceives is far from veridical and complete. As a result, the existing beliefs of an agent may conflict with some new information. So, belief revision is necessary for a rational agent to preserve the consistency of her knowledge. Related to belief revision, decision-making is a form of reasoning to determine actions. Given some available information about the current status of the world and the consequences of potential actions, there could be more than one alternative [2]. Since the available information may be incomplete and uncertain and different alternatives could not be selected simultaneously, conflicts may arise in the process of decision-making. For instance, in medical reasoning, knowledge can be uncertain and inconsistent [3], and more than one explanation for a medical hypothesis could be possible based on different medical studies [4]; in norm-governed systems, an agent may have several conflicting motivations, including internal desires and external obligations that are brought about by norms and policies [5].

In multi-agent systems, different agents interact with each other to achieve certain goals. When agents have competing claims on scarce resources (such as commodities, services, time, money, etc.), not all of them can be simultaneously satisfied. First, for those agents that have

Efficient Computation of Argumentation Semantics. http://dx.doi.org/10.1016/B978-0-12-410406-8.00001-4

conflicting interests and a desire to cooperate, negotiation is an important coordination mechanism [6]. Due to the conflicting interests and uncertain information from the environment, how to formulate the mechanisms of negotiation is still an open problem. Second, for those agents that have opposite goals, argumentation is a form of reasoning for each agent to pursue her own goals. Legal argumentation is a typical example. It is adversarial and dialectic in nature [7]. Since argumentation schemes (a kind of generalized rules of inference) need not concern strict, abstract, necessarily valid patterns of reasoning, but can be defeasible, concrete and contingently valid, the conclusions of different schemes could be conflicting and defeasible [8].

In addition, for some practical reasoning systems in open and dynamic environments, such as semantic web [9] and trust computing [10], inconsistencies of information are also very common. For instance, in semantic web, domain ontologies established by different parties may conflict. In trust computing, both the information and its sources are uncertain. The conflicts among them are inevitable.

In order to treat with the uncertainty, incompleteness, and inconsistency of information, a number of research efforts have been made in the area of non-monotonic reasoning. In the 1980s, three non-monotonic formalisms (including default logic [11], circumscription [12] and autoepistemic logic [13]) were proposed. However, these formalisms are mainly suitable for epistemic reasoning. For practical reasoning and the reasoning in the process of multi-agent interaction, a more general non-monotonic formalism is needed. Meanwhile, from the perspective of implementation, the computational complexity problems of these formalisms are challenging. Furthermore, when considering the dynamics of systems (when new information is added to a system), the efficiency of computation is a great problem. This is because when a system is updated, any conclusions obtained previously could be overruled by some counter-arguments. So, for each modification, all existing conclusions should be recomputed. This is obviously inefficient.

1.2 The Notion of Argumentation

Due to the above-mentioned problems, in the past fifteen years, argumentation has been recognized as a non-monotonic formalism that is more general and natural than the traditional ones. In general, the study of argumentation may, informally, be considered as concerned with how assertions are proposed, discussed, and resolved in the context of disagreement [14].

The disagreement (inconsistency) may arise during the process of reasoning of an individual agent, or a set of agents interacting with each other. In different cases, the nature of inconsistency may vary. As mentioned above, in the process of epistemic reasoning

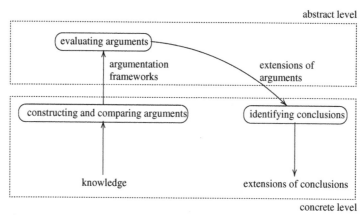

Figure 1.1 The working mechanism of an argumentation system.

(belief revision), the inconsistency of information is mainly due to the uncertainty and incompleteness of information. In the case of practical reasoning (decision-making, planning, etc.), an agent may have several motivations (desires, obligations, etc.). Due to the limitation of resources, the agent can not fulfil all of them, and the conflicts among different motivations arise. In means-end reasoning, there exist different options, which could be conflicting. In the case of inter-agent communication, such as negotiation and discussion, the interests, objectives, preferences or standpoints of different participants might be inconsistent.

Despite the fact that various arguments may have different internal structures and the nature of inconsistence between different kinds of arguments may vary, when evaluating the status of arguments, we may regard the attacks between arguments as abstract binary relations. Based on this idea, in 1995, Phan Ming Dung proposed an abstract argumentation theory, in which arguments are abstract entities, while the origins of attacks are also left unknown [15]. As a result, an argument system is simply represented as a directed graph (called a defeat graph), represented as $F = (A, R)$, where A denotes a set of arguments and $R \subseteq A \times A$ denotes a set of attacks.

Since the status of arguments is evaluated in an abstract argumentation framework, while the representation of underlying knowledge, the construction of arguments and the identification of attacks are comparatively independent, an argumentation system can be conceived of a two-level architecture (Figure 1.1.), and its working process is composed of three steps. First, according to a set of underlying knowledge, a set of arguments are constructed and the attacks between them are identified, forming an argumentation framework. Second, given an argumentation framework, the status of arguments is evaluated (in terms of certain criteria), producing sets of extensions of arguments (each extension could be understood as a set of arguments that are acceptable together). Third, for each extension of arguments, the associated set of conclusions is identified.

Let us consider the following example.

Example 1.1. The following is a story, called "Two Children Arguing about the Sun".

When Confucius was traveling in the eastern part of the country, he came upon two children hot in argument, so he asked them to tell him what it was all about.

- "I think," said one child, "that the sun is near to us at daybreak and far away from us at noon."
- The other contended that the sun was far away at dawn and nearby at midday.
- "When the sun first appears," said one child, "it is as big as the canopy of a carriage, but at noon it is only the size of a plate or a bowl. Well, isn't it true that objects far away seem smaller while those nearby seem bigger?"
- "When the sun comes out," pointed out the other, "it is very cool, but at midday it is as hot as putting your hand in boiling water. Well, isn't it true that what is nearer to us is hotter and what is farther off is cooler?"

Confucius was unable to settle the matter for them. The two children laughed at him, "Who says you are a learned man?"

Now, let a denote the argument raised by one child, and b the argument raised by the other. Then, we get an argumentation framework $F = (\{a, b\}, \{(a, b), (b, a)\})$. For a skeptical agent, he accepts neither a nor b, i.e., the set (extension) of arguments is empty. Note that in this example, the underlying logic for representing the knowledge and the methods for constructing the arguments and identifying the attacks are independent of the evaluation of the arguments.

Since abstract argumentation is comparatively independent of the underlying argument construction and comparison, it shows great promise as a theoretical tool for handling the various reasoning problems under the context of inconsistency. It is not only suitable for the reasoning of individual agents, but also for the interactions of multiple agents. As a result, argumentation is now regarded as a non-monotonic formalism for logical deduction, induction, abduction and plausible reasoning, and is promising to be applied to various applications, such as semantic web, recommender systems, trust computing, normative systems, social choice theory, judgment aggregation, law and medical, etc.

1.3 Motivations of this Book

Given an argumentation framework $F = (A, R)$, a basic problem is to determine which arguments can be regarded as acceptable together. In order to resolve this problem, a lot of evaluating criteria (often called argumentation semantics) have been proposed. Some typical argumentation semantics include admissible, complete, preferred, grounded, stable, semi-stable, ideal and eager, etc. Each semantics stands for a set of criteria. For instance, a set of arguments $B \subseteq A$ is regarded as *admissible*, if it is *conflict-free* and it is able to *defend* each argument in it. Informally, we say that B is conflict-free, it there exist no arguments $a, b \in A$ such that a attacks b; an argument a is defended by B, if for every argument $b \in A$ that attacks a then there exists an argument $c \in B$ such that c attacks b. Definitions of some other semantics will be presented in the next chapter.

Under many semantics, determining the status of arguments is often not easy. It has been proved that many natural questions regarding argument acceptability are, in general, computationally intractable [16]. For this reason, in recent years, an increasing amount of work has been done on identifying tractable fragments of argumentation frameworks and developing more efficient algorithms [17,18]. However, how to efficiently compute the semantics of a generic argumentation framework is still an open problem.

In order to efficiently compute the semantics of a generic argumentation framework, the following two lines of work have been undertaken. The first line of work is on developing efficient approaches by exploiting existing sophisticated problem solvers that have been developed for other purposes. According to existing literature, the approaches based on propositional logic [19–22] and answer-set programming [23–25], respectively, have been widely studied. They are also called *reduction-based approaches*, which reduce an abstract argumentation problem into an instance of a different problem (SAT or ASP), delegating the burden of computation to existing systems. Thanks to the efficiency of the solvers for SAT or ASP problems, the corresponding argumentation problems could be resolved more efficiently. The second line of work is on establishing efficient approaches dedicated to argumentation, including labelling-based approaches [26–28] and dialogue games [26,29]. These approaches implement procedures for abstract argumentation from scratch, and therefore are called *direct approaches*. By adopting some specific techniques, the efficiency of direct approaches could be improved to some extent [27].

Although the above-mentioned two lines of work have made great efforts to improve the efficiency of computation, they all treat each argumentation framework as a monolithic entity, which may fundamentally restrict the improvement of efficiency. On the contrary, if we compute the status of arguments locally, then the computation may become easier. Let us consider the following three examples.

First, for most argumentation systems, changing of arguments and their attack relation is an intrinsic property of them. When an argumentation system is modified, if we recompute the status of each argument afresh, it is obviously inefficient. As a matter of fact, for each modification to an argumentation system, it is very common that only a part of arguments in the system are affected. Hence, if it is feasible to evaluate the status of those affected arguments locally, then the remaining arguments are not necessary to be considered.

Second, according to [16], although for some argumentation frameworks with special structures (such as acyclic, symmetric and bounded tree-width) there exist tractable algorithms, for a generic argumentation framework that might not belong to any tractable class, how to efficiently compute its semantics still remains unresolved. A promising solution is to decompose a general argumentation framework into a set of sub-frameworks, such that the status of each argument in a sub-framework could be computed locally. Given that the status of each argument is computed in a sub-framework, which might belong to some tractable class and has smaller size than that of the original argumentation framework, the execution time for evaluating the status of arguments might decrease.

Third, when querying the status of some arguments in an argumentation framework, we may only take into consideration a subset of arguments in the argumentation framework that are relevant to these arguments. In other words, the computation of the status of some arguments is performed locally.

From the above three examples, we observe that local computation could be an effective means to decrease the computation time in many situations. This is due to the following two reasons. In the first place, compared to the size of a whole argumentation framework, the size of a sub-framework is often smaller or much smaller. This idea has been adopted in divide-and-conquer algorithms, which have been recognized as an important algorithm design paradigm. In the second place, when an argumentation framework does not belong to any tractable class, some of its sub-frameworks might belong to some tractable classes.

Inspired by the above ideas, this book will introduce efficient approaches for computing the semantics of argumentation based on the theory of local computation.

1.4 The Structure of this Book

This book focuses on formulating efficient approaches for computing the semantics of abstract argumentation frameworks, and is divided into five parts. Besides this part for introduction, the contents of other four parts are as follows.

In Part II, some fundamental theories of argumentation as well as some existing approaches and algorithms for computing argumentation semantics are presented. For readers who are familiar with the area of computational argumentation, this part may be skipped.

In Part III, we introduce a theory of computation based on the idea of local computation. This part consists of two chapters. In Chapter 4, we first introduce the notion and the semantics of sub-frameworks. In Chapter 5, we formulate the relations between the semantics of an argumentation framework (global semantics) and that of a sub-framework (local semantics). This part lays a foundation for the subsequent part.

In Part IV, we formulate three approaches for efficient computation of argumentation from different perspectives. Specifically, Chapter 6 deals with the computation of the status of arguments in a static argumentation framework, Chapter 7 studies the computation of semantics of a dynamic argumentation system, and Chapter 8 deals with the computation of a part of arguments in an argumentation framework.

In Part V, we conclude the book and point out some future work.

References

[1] J.L. Pollock, Perceiving and reasoning about a changing world, Computational Intelligence 14 (4) (1998) 498–562.

[2] L. Amgoud, H. Prade, Using arguments for making and explaining decisions, Artificial Intelligence 173 (2009) 413–436.

[3] J. Fox, D. Glasspool, D. Grecu, S. Modgil, M. South, V. Patkar, Argumentation-based inference and decision making–a medical perspective, IEEE Intelligent Systems 22 (6) (2007) 34–40.

[4] M.A. Grando, L. Moss, D. Sleeman, J. Kinsella, Argumentation-logic for creating and explaining medical hypotheses, Artificial Intelligence in Medicine 58 (1) (2013) 1–13.

[5] B. Liao, H. Huang, ANGLE: an autonomous, normative and guidable agent with changing knowledge, Information Sciences 180 (17) (2010) 3117–3139.

[6] I. Rahwan, S.D. Ramchurn, N.R. Jennings, P. McBurney, S. Parsons, L. Sonenberg, Argumentation-based negotiation, The Knowledge Engineering Review 18 (4) (2003) 343–375.

[7] T. Bench-Capon, H. Prakken, G. Sartor, Argumentation in legal reasoning, in: I. Rahwan, G.R. Simari (Eds.), Argumentation in Artificial Intelligence, Springer, 2009, pp. 363–382.

[8] B. Verheij, Dialectical argumentation with argumentation schemes: an approach to legal logic, Artificial Intelligence and Law 11 (2–3) (2003) 167–195.

[9] S.A. Gómez, C.I. Chesñevar, G.R. Simari, ONTOarg: a decision support framework for ontology integration based on argumentation, Expert Systems with Applications 40 (5) (2013) 1858–1870.

[10] S. Villata, G. Boella, D.M. Gabbay, L. van der Torre, A socio-cognitive model of trust using argumentation theory, International Journal of Approximate Reasoning 54 (4) (2013) 541–559.

[11] R. Reiter, A logic for default reasoning, Artificial Intelligence 13 (1–2) (1980) 81–132.

[12] J. McCarthy, Circumscription–a form of non-monotonic reasoning, Artificial Intelligence 13 (1–2) (1980) 27–39.

[13] R.C. Moore, Semantic considerations on nonmonotonic logic, Artificial Intelligence 25 (1) (1985) 75–94.

[14] T.J.M. Bench-Capon, P.E. Dunne, Argumentation in artificial intelligence, Artificial Intelligence 171 (10–15) (2007) 619–641.

[15] P.M. Dung, On the acceptability of arguments and its fundamental role in nonmonotonic reasoning, logic programming and n-person games, Artificial Intelligence 77 (2) (1995) 321–357.

[16] P.E. Dunne, Computational properties of argument systems satisfying graph-theoretic constraints, Artificial Intelligence 171 (10–15) (2007) 701–729.

[17] S. Coste-Marquis, C. Devred, P. Marquis, Symmetric argumentation frameworks, in: Proceedings of the Eighth European Conference Symbolic and Quantitative Approaches to Reasoning with Uncertainty, 2005, pp. 317–328.

[18] W. Dvořák, R. Pichler, S. Woltran, Towards fixed-parameter tractable algorithms for argumentation, in: Proceedings of the 12th International Conference on the Principles of Knowledge Representation and Reasoning, 2010, pp. 112–122.

[19] O. Arieli, M. Caminada, A general QBF-based formalization of abstract argumentation theory, in: Proceedings of the Fourth Conference on Computational Models of Argument, 2012, pp. 105–116.

[20] P. Besnard, S. Doutre, Checking the acceptability of a set of arguments, in: Proceedings of the 10th International Workshop on Non-Monotonic Reasoning, 2004, pp. 59–64.

[21] W. Dvořák, M. Järvisalo, J.P. Wallner, S. Woltran, Complexity-sensitive decision procedures for abstract argumentation, in: Proceedings of the 13th International Conference on Principles of Knowledge Representation and Reasoning, 2012, pp. 54–64.

[22] U. Egly, S. Woltran, Reasoning in argumentation frameworks using quantified boolean formulas, in: Proceedings of the First Conference on Computational Models of Argument, 2006, pp. 133–144.

[23] U. Egly, S.A. Gaggl, S. Woltran, Answer-set programming encodings for argumentation frameworks, Argument & Computation 1 (2) (2010) 147–177.

[24] J.C. Nieves, M. Osorio, U. Cortés, Preferred extensions as stable models, Theory and Practice of Logic Programming 8 (4) (2008) 527–543.

[25] T. Wakaki, K. Nitta, Computing argumentation semantics in answer set programming, in: Proceedings of the 22nd Annual Conference of the Japanese Society for Artificial Intelligence, 2008, pp. 254–269.

[26] S. Modgil, M. Caminada, Proof theories and algorithms for abstract argumentation frameworks, in: I. Rahwan, G.R. Simari (Eds.), Argumentation in Artificial Intelligence, Springer, 2009, pp. 105–132.

[27] S. Nofal, P.E. Dunne, K. Atkinson, On preferred extension enumeration in abstract argumentation, in: Proceedings of the Fourth Conference on Computational Models of Argument, 2012, pp. 205–216.

[28] B. Verheij, A labeling approach to the computation of credulous acceptance in argumentation, in: Proceedings of the 20th International Joint Conference on Artificial Intelligence, 2007, pp. 623–628.

[29] P.M. Thang, P.M. Dung, N.D. Hung, Towards a common framework for dialectical proof procedures in abstract argumentation, Journal of Logic and Computation 19 (6) (2009) 1071–1109.

CHAPTER 2

Semantics of Argumentation

Chapter Outline

2.1 Introduction

Given a set of conflicting arguments, whether an argument can be accepted depends on the existence of some counter arguments, which may in turn have counter arguments, and so on. In order to capture the attack relation between arguments and facilitate the status evaluation of arguments, Phan Minh Dung proposed abstract argumentation theory in 1995 [1]. He introduced a notion of abstract argumentation framework. An abstract argumentation framework can be regarded as a directed graph where the nodes represent arguments and the arcs represent attacks. Given such a graph, a fundamental problem is to determine which arguments can be regarded as acceptable. According to existing literature, there are mainly two approaches to deal with this problem: extension-based approach and labelling-based

Efficient Computation of Argumentation Semantics. http://dx.doi.org/10.1016/B978-0-12-410406-8.00002-6

approach. Both approaches aim at defining argumentation semantics. An argumentation semantics can be viewed as a pre-defined criterion, according to which the acceptability of arguments in an argumentation framework can be determined.

In this chapter, after the notion of an abstract argumentation framework is introduced, some mainstream semantics defined by the two approaches are formulated.

2.2 Abstract Argumentation Frameworks

The notion of abstract argumentation frameworks (briefly, argumentation frameworks, or AFs) is central to abstract theory of argumentation. As mentioned above, in order to reflect the conflicting relation of arguments and facilitate the status evaluation of arguments, the structures of arguments and the origins of attacks that are irrelevant to the status evaluation of arguments are omitted. As a result, an argumentation framework is simply represented as a directed graph (called *a defeat graph*) where nodes represent arguments and arcs represent attacks between them. Formally, an argumentation framework is defined as follows.

Definition 2.1 (Abstract argumentation framework).   An abstract argumentation framework is a tuple $F = (A, R)$, where A is a set of arguments, and $R \subseteq A \times A$ is a set of attacks.

In Definition 2.1, we use $(\alpha, \beta) \in R$ to denote that α attacks β (with respect to R). Meanwhile, throughout this book, *we assume that A is generated by a reasoner at a given time point, and therefore is finite.*

Let $B \subseteq A$ be a set of arguments, and $\alpha \in A$ be an argument. As usual, we say that α attacks B if and only if there exists $\beta \in B$ such that α attacks β. Meanwhile, we say that B attacks α if and only if there exists $\beta \in B$ such that β attacks α.

In a defeat graph, the nodes that attack a given argument α are called *defeaters* (or *parents*) of α [2]. Given $F = (A, R)$ and a set $B \subseteq A$ of arguments, we use B^- to denote the set of *outside parents* of the arguments in B (when $B^- = \emptyset$, B is called an *unattacked set*).

$$B^- = \{\alpha \in A \setminus B \mid \exists \beta \in B, \quad \text{such that } (\alpha, \beta) \in R\} \tag{2.1}$$

2.3 Argumentation Semantics

Since there exist conflicts between arguments, to determine whether or not an argument is acceptable, the status of all its defeaters should be considered, while the status of each defeater is in turn determined by the status of its defeaters, and so on. When an argument α is attacked by an argument β, and β is in turn attacked by an argument γ, then we say that γ reinstates α. Whether an argument is able to reinstate another argument (or itself) is related to some specific criterion, which is called *argumentation semantics*. Given an argumentation framework, under a certain argumentation semantics, zero or more sets of arguments are considered acceptable.

In existing literature, there are two approaches to formulate the status of arguments in an argumentation framework: extension-based approach and labelling-based approach. The idea underlying the extension-based approach is to identify sets of arguments, called extensions, that can be regarded as collectively acceptable according to some criterion. The idea underlying the labelling-based approach is to assign a label to each argument, according to a certain criterion.

In this chapter, we will introduce some basic notions related to argumentation semantics, defined by the above-mentioned two approaches. Readers may refer to [3] for an excellent introduction.

2.3.1 Extension-based Approach

In the extension-based approach, under different semantics which represent different evaluation criteria, an argumentation framework may have different sets of extensions. In this section, the following semantics will be introduced: admissible, complete, grounded, preferred, stable, semi-stable, eager and ideal.

2.3.1.1 Admissible Extension

The notion of admissible is fundamental to the definitions of other semantics. It is defined on the basis of the following two notions: *conflict-free* and *acceptable*.

Conflict-freeness is viewed as a minimal requirement of any argumentation semantics.

Definition 2.2 (Conflict-free). Let $F = (A, R)$ be an argumentation framework, and $B \subseteq A$ a set of arguments. B is *conflict-free* if and only if $\nexists \alpha, \beta \in B$, such that $(\alpha, \beta) \in R$.

However, conflict-freeness is too weak a condition to justify that a set of arguments is "collectively acceptable", since such a set could be attacked by arguments not among its members.

Example 2.1. Let $F_{2.1}$ be an argumentation framework (Figure 2.1). The following extensions are conflict-free.

- $\emptyset$;
- $\{a\}, \{b\}, \{c\}$;
- $\{a, c\}$.

$$a \longrightarrow b \longrightarrow c$$

Figure 2.1 Argumentation framework $F_{2.1}$.

However, if we require that a set of collectively acceptable arguments should be able to defend itself, then not all of these subsets satisfy the requirement. For instance, the set $\{c\}$ can not defend itself, because Argument c is attacked by Argument b, and there exists no argument in $\{c\}$ that attacks b.

Acceptability is another important requirement for the arguments that constitute an admissible extension.

Definition 2.3 (Acceptable).   Let $F = (A, R)$ be an argumentation framework, and $B \subseteq A$ a set of arguments. An argument $\alpha \in A$ is *acceptable* with respect to B (also called α is *defended* by B), if and only if $\forall (\beta, \alpha) \in R$, $\exists \gamma \in B$, such that $(\gamma, \beta) \in R$.

Based on the notions of conflict-freeness and acceptability, an *admissible extension* can be defined as follows.

Definition 2.4 (Admissible extension).   Let $F = (A, R)$ be an argumentation framework, and $B \subseteq A$ a set of arguments. B is *admissible* if and only if B is conflict-free, and each argument in B is acceptable with respect to B.

Example 2.2.   Continue Example 2.1. According to Definition 2.4, the following extensions are admissible:

- $\emptyset$;
- $\{a\}$;
- $\{a, c\}$.

2.3.1.2 Complete Extension

As illustrated in Example 2.2, it is not the case that each admissible extension contains all arguments that are acceptable with respect to the extension. For instance, with respect to $\{a\}$, the argument c is acceptable, but it is not in $\{a\}$.

Given an argumentation framework $F = (A, R)$ and an admissible extension $B \subseteq A$, if there exists an argument $\alpha \in A \setminus B$ such that α is acceptable with respect to B, then we add α to B, and get a new set $B' = B \cup \{\alpha\}$. Then, with respect to B', if there exists an argument $\beta \in A \setminus B'$ such that α is acceptable with respect to B', then we add β to B', and get a new set $B'' = B' \cup \{\beta\}$. In this way, if A is a finite set, then we will finally get a set that is admissible and contains all arguments that are acceptable with respect to this set. This process can be defined by a function, called *characteristic function*, defined as follows.

Definition 2.5 (Characteristic function).   Let $F = (A, R)$ be an argumentation framework. The *characteristic function* of F, denoted as $\mathscr{F}_F$, is defined as follows:

$$\mathscr{F}_F : 2^A \rightarrow 2^A,$$

$\mathscr{F}_F(S) = \{\alpha \mid \alpha$ is acceptable with respect to $S\}$.

Given an admissible set B, if $B = \mathscr{F}_F(B)$, then B is called a *complete extension*. Formally, we have the following definition.

Definition 2.6 (Complete extension). Let $F = (A, R)$ be an argumentation framework. We say that $B \subseteq A$ is a complete extension if and only if B is admissible and each argument in A that is acceptable with respect to B is in B, i.e., $B = \mathscr{F}_F(B)$.

Example 2.3. Continue Example 2.2. According to Definition 2.6, only $\{a, c\}$ is a complete extension.

Under complete semantics, some argumentation frameworks may have more than one extension. Let us consider the following example.

Example 2.4. Let $F_{2.2}$ be an argumentation framework (Figure 2.2). It has three complete extensions as follows:

- $\{a, c, d\}$;
- $\{b, d\}$;
- $\{d\}$.

2.3.1.3 Grounded Extension and Preferred Extension

As mentioned above, under complete semantics, there might exist several extensions. The arguments in different extensions might be conflicting, and therefore questionable. If we only accept those arguments that are least questionable, then we get an extension (called *grounded extension*) that is the most skeptical among all complete extensions. On the contrary, if we want to accept as many arguments as reasonably possible, then we get sets of extensions (called *preferred extensions*) that are more credulous than some other complete extensions.

According to Example 2.4, the argument d is acceptable with respect to all extensions, while the arguments a, b and c are not. So, under grounded semantics, one may only regard d as an acceptable argument, while under preferred semantics, one may regard a, c and d (or b and d) as acceptable arguments.

Definition 2.7 (Grounded extension and preferred extension). Let $F = (A, R)$ be an argumentation framework, and $B \subseteq A$ be a set of arguments.

$$a \rightleftharpoons b \longrightarrow c \qquad d$$

Figure 2.2 Argumentation framework $F_{2.2}$.

- B is a grounded extension if and only if B is the minimal (with respect to set-inclusion) complete extension.
- B is a preferred extension if and only if B is a maximal (with respect to set-inclusion) complete extension.

According to [1], for any argumentation framework, there exists a unique grounded extension, while for some argumentation frameworks, there might exist multiple preferred extensions.

Example 2.5.   Continue Example 2.4. According to Definition 2.7, it holds that

- $\{d\}$ is the grounded extension;
- $\{a, c, d\}$ and $\{b, d\}$ are preferred extensions.

Grounded extension can be obtained by recursively applying characteristic function from an empty set.

Example 2.6.   Consider the argumentation framework in Example 2.1. Let $S = \emptyset$. According to Definition 2.5, it holds that

- $\mathscr{F}_{F_{2.1}}(S) = \{a\}$;
- $\mathscr{F}^2_{F_{2.1}}(S) = \{a, c\}$;
- $\mathscr{F}^3_{F_{2.1}}(S) = \mathscr{F}^2_{F_{2.1}}(S)$.

Hence, the grounded extension of $F_{2.1}$ is $\{a, c\}$.

Alternatively, the grounded extension of an argumentation framework can be defined as follows.

Definition 2.8 (Grounded extension).   Let $F = (A, R)$ be an argumentation framework, and $B \subseteq A$ be a set of arguments. B is a grounded extension if and only if B is the minimal (with respect to set-inclusion) conflict-free fixed point of the characteristic function $\mathscr{F}_F$.

2.3.1.4 Stable Extension and Semi-Stable Extension

Under the above-mentioned semantics, with respect to a given extension, the status of some arguments could be *undecided*, which means that they are neither *accepted*, nor *rejected*, with respect to the extension. Here, we say that an argument is accepted with respect to an extension, if it belongs to this extension; an argument is rejected with respect to an extension if it is attacked by the extension.

Example 2.7.   Let $F_{2.3}$ be an argumentation framework (Figure 2.3). Under admissible, complete, grounded, or preferred semantics, it has only one extension $\{a\}$. The status of arguments c, d and e is undecided.

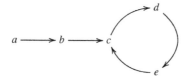

Figure 2.3   Argumentation framework $F_{2.3}$.

Now, given a set of conflict-free arguments, if we require that there are no arguments that are undecided with respect to this set, then it is called a stable extension. Formally, we have the following definition.

Definition 2.9 (Stable extension).   Let $F = (A, R)$ be an argumentation framework, and $B \subseteq A$ be a set of arguments. B is a stable extension if and only if B is conflict-free and B attacks each argument that does not belong to B.

In [1], it has been proved that every stable extension is a preferred extension, but not vice versa.

Stable semantics is interesting, since it exactly corresponds to the extensions defined in some traditional non-monotonic formalisms, such as Reiter's default logic, Moore's autoepistemic logic and logic programming. However, some argumentation frameworks may have no stable extension.

Example 2.8.   Continue Example 2.7. Argumentation framework $F_{2.3}$ has no stable extension.

Since under stable semantics there is a possibility that stable extensions may not exist, Martin Caminada proposed a revised semantics, called semi-stable semantics. This semantics is "backward compatible"to stable semantics in the sense that it is equivalent to stable semantics in situations where stable extensions exist, and still yields a reasonable result in situations where stable extensions do not exist [4]. Compared to stable semantics, semi-stable semantics does not require that the set of undecided arguments is empty, but merely requires that the set of undecided arguments is minimal.

Definition 2.10 (Semi-stable extension).   Let $F = (A, R)$ be an argumentation framework, and $B \subseteq A$ be a set of arguments. B is a semi-stable extension if and only if B is a complete extension of which $B \cup B^+$ is maximal, in which $B^+ = \{\alpha \in A \mid \exists \beta \in B$ such that $(\beta, \alpha) \in R\}$.

It has been verified that every stable extension is a semi-stable extension, and every semi-stable extension is a preferred extension.

Example 2.9.   Continue Example 2.7. Argumentation framework $F_{2.3}$ has a semi-stable extension $\{a\}$.

2.3.1.5 Ideal Extension and Eager Extension

As illustrated in Example 2.5, an argumentation framework may have several preferred extensions, and some arguments in different extensions might conflict. More specifically, a and b are respectively in $\{a, c, d\}$ and $\{b, d\}$, but they are conflicting. This is because preferred semantics is credulous. In many cases, it is better to adopt a semantics that is more skeptical. The above-mentioned grounded semantics is skeptical. However, it is often overly skeptical, in that in many cases, the grounded extension could be an empty set. In order to treat this problem, another skeptical semantics, called *ideal* semantics, was proposed in [5]. It defines an ideal extension as an admissible extension that is a subset of every preferred extension. Generally, ideal semantics is less skeptical than grounded semantics, but more skeptical than preferred semantics. Formally, we have the following definition.

Definition 2.11 (Ideal extension).   Let $F = (A, R)$ be an argumentation framework, and $B \subseteq A$ be a set of arguments. B is ideal if and only if B is admissible and it is contained in every preferred set of arguments. The ideal extension is the maximal (with respect to set-inclusion) ideal set.

Example 2.10.   Continue Example 2.5. According to Definition 2.11, $\{d\}$ is the ideal extension. It is also the grounded extension, and the intersection of the two preferred extensions.

Example 2.10 shows that the ideal extension is the same as the grounded extension. Meanwhile, the intersection of the preferred extensions is equal to the ideal extension. However, this coincidence does not happen in all cases. Please consider the following example (Figure 2.4).

Example 2.11.   According to the above definition, we have

- $\{\}$ is the grounded extension;
- $\{a, c, f, j\}, \{a, d, f, j\}, \{a, c, g, j\}$ and $\{a, d, g, j\}$ are four preferred extensions.
- $\{a\}$ is the ideal extension.

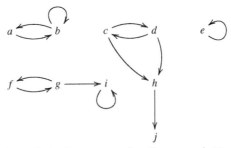

Figure 2.4   Argumentation framework $F_{2.4}$.

In this example, the grounded extension is not equal to the ideal extension. Meanwhile, since the intersection of the two preferred extensions $\{a, j\}$ is not admissible, it is not an ideal set.

Besides grounded semantics and ideal semantics, the third skeptical semantics that is very close to ideal semantics is eager semantics. While the ideal extension is the biggest admissible subset of each preferred extension, the eager extension is the biggest admissible subset of each semi-stable extension.

Example 2.12. Continue Example 2.11.

- $\{a, c, g, j\}$ and $\{a, d, g, j\}$ are two semi-stable extensions.
- $\{a, g\}$ is the eager extension.

2.3.2 Labelling-based Approach

In labelling-based approaches, there are usually three labels: IN, OUT and UNDEC, where the label IN means the argument is accepted, the label OUT means the argument is rejected and the label UNDEC means one abstains from an opinion on whether the argument is accepted or rejected [3]. Meanwhile, there could be some other choices for the set of labels. For instance, in [6], four-valued labelling is considered. In this book, we choose three-valued labelling. Formally, labelling is defined as follows [7].

Definition 2.12 (Labelling). Given an argumentation framework $F = (A, R)$ and three labels IN, OUT and UNDEC, a *labelling* is defined as a total function

$$\mathcal{L} : A \mapsto \{\text{IN, OUT, UNDEC}\}$$

The set of all labellings of F is denoted as Lab_F.

Let $in(\mathcal{L}) = \{\alpha \mid \mathcal{L}(\alpha) = \text{IN}\}$, $out(\mathcal{L}) = \{\alpha \mid \mathcal{L}(\alpha) = \text{OUT}\}$, and $undec(\mathcal{L}) = \{\alpha \mid \mathcal{L}(\alpha) = \text{UNDEC}\}$. A labelling $\mathcal{L}$ is often represented as a triple of the form $(in(\mathcal{L}), out(\mathcal{L}), undec(\mathcal{L}))$.

One of the criteria for labelling-based semantics is whether a label assigned to an argument is legal. According to Definition 2.12, given a labelling $\mathcal{L}$, the status assigned to each argument might not be legal. We say that assigning the IN label to an argument α is legal if and only if all its attackers have been assigned the OUT label; assigning the OUT label to an argument α is legal if and only if one of its attackers has been assigned the IN label; and assigning the UNDEC label to an argument α is legal if and only if not all its attacks are labelled OUT and it does not have an attacker that is labelled IN. Based on [8], we have the following definition.

Definition 2.13 (Legal labelling). Let $\mathcal{L}$ be a labelling of an argumentation framework $F = (A, R)$ and $\alpha \in A$.

- α is legally IN if and only if $\mathcal{L}(\alpha) = $ IN and for all $\beta \in A$, if $(\beta, \alpha) \in R$, then $\mathcal{L}(\beta) = $ OUT.
- α is legally OUT if and only if $\mathcal{L}(\alpha) = $ OUT and there exists $\beta \in A$, such that $(\beta, \alpha) \in R$, and $\mathcal{L}(\beta) = $ IN.
- α is legally UNDEC if and only if $\mathcal{L}(\alpha) = $ UNDEC and

  (1) for all $\beta \in A$, if $(\beta, \alpha) \in R$, then $\mathcal{L}(\beta) \neq $ IN, and
  (2) it is not the case that: for all $\beta \in A$, if $(\beta, \alpha) \in R$, then $\mathcal{L}(\beta) = $ OUT.

According to the notion of legal labelling, the notion of illegal labelling can be defined as follows.

Definition 2.14 (Illegal labelling).   Let $\mathcal{L}$ be a labelling of an argumentation framework $F = (A, R)$ and $\alpha \in A$. For $l \in \{$IN, OUT, UNDEC$\}$, α is illegally l if and only if $\mathcal{L}(\alpha) = l$, but α is not legally l.

Based on the notions of labelling as well as legal/illegal labelling, various labelling-based semantics corresponding to the above-mentioned extension-based semantics can be defined as follows.

2.3.2.1 Admissible Labelling

Admissible labeling is defined on the basis of conflict-free labelling. According to Definition 2.2, when a set of arguments is conflict-free, there exists no attack between the arguments in the set. So, given a labelling $\mathcal{L}$, each argument in $in(\mathcal{L})$ should not have an attacker in $in(\mathcal{L})$. Meanwhile, only those arguments that are attacked by at least one argument in $in(\mathcal{L})$ are labelled OUT. So, all OUT-labelled arguments are legally OUT.

Definition 2.15 (Conflict-free labelling).   Let $F = (A, R)$ be an argumentation framework and $\mathcal{L} : A \mapsto \{$IN, OUT, UNDEC$\}$ be a total function. $\mathcal{L}$ is a conflict-free labelling, if and only if each argument that is labelled IN does not have an attacker that is labelled IN, and each argument that is labelled OUT is legally OUT.

Example 2.13.   Continue Example 2.1. With respect to $F_{2.1}$, the following labellings are conflict-free:

- $(\emptyset, \emptyset, \{a, b, c\})$
- $(\{a\}, \{b\}, \{c\})$
- $(\{a\}, \emptyset, \{b, c\})$
- $(\{b\}, \{c\}, \{a\})$
- $(\{b\}, \emptyset, \{a, c\})$
- $(\{c\}, \emptyset, \{a, b\})$
- $(\{a, c\}, \{b\}, \emptyset)$
- $(\{a, c\}, \emptyset, \{b\})$

From this example, we may observe that a conflict-free set might correspond to several conflict-free labellings.

According to Definition 2.15, not all IN-labelled arguments are legally IN. Given a conflict-free labelling, if each IN-labelled argument is legally IN, then it is called an admissible labelling.

Definition 2.16 (Admissible labelling). Let $F = (A, R)$ be an argumentation framework and $\mathfrak{L} : A \mapsto \{$IN, OUT, UNDEC$\}$ be a total function. $\mathfrak{L}$ is an admissible labelling, if and only if it is a conflict-free labelling, and each argument that is labelled IN is legally IN.

Example 2.14. Continue Example 2.13. According to Definition 2.16, the following labellings are admissible labellings:

- $(\emptyset, \emptyset, \{a, b, c\})$
- $(\{a\}, \{b\}, \{c\})$
- $(\{a\}, \emptyset, \{b, c\})$
- $(\{a, c\}, \{b\}, \emptyset)$
- $(\{a, c\}, \emptyset, \{b\})$

2.3.2.2 Complete Labelling

Given an admissible labelling, if each UNDEC-labelled argument is legally UNDEC, then it is called a complete labelling. Formally, we have the following definition.

Definition 2.17 (Complete labelling). Let $F = (A, R)$ be an argumentation framework and $\mathfrak{L} : A \mapsto \{$IN, OUT, UNDEC$\}$ be a total function. $\mathfrak{L}$ is a complete labelling, if and only if it is an admissible labelling, and each argument that is labelled UNDEC is legally UNDEC.

Example 2.15. Continue Example 2.14. According to Definition 2.17, only $(\{a, c\}, \{b\}, \emptyset)$ is a complete labelling.

2.3.2.3 Grounded Labelling and Preferred Labelling

Given a complete labelling, if the set of IN-labelled arguments is minimal (respectively, maximal), then it is called a grounded labelling (respectively, preferred labelling). Formally, we have the following definition.

Definition 2.18 (Grounded labelling and preferred labelling). Let $F = (A, R)$ be an argumentation framework and $\mathfrak{L} : A \mapsto \{$IN, OUT, UNDEC$\}$ be a total function.

- $\mathfrak{L}$ is a grounded labelling, if and only if it is a complete labelling, and $in(\mathfrak{L})$ is minimal (with respect to set-inclusion).
- $\mathfrak{L}$ is a preferred labelling, if and only if it is a complete labelling, and $in(\mathfrak{L})$ is maximal (with respect to set-inclusion).

Example 2.16. Consider the argumentation framework in Example 2.11. According to Definition 2.18, it holds that

- $(\emptyset, \emptyset, \{a, b, c, d, e, f, g, h, i, j\})$ is the grounded labelling;
- $(\{a, c, f, j\}, \{b, d, h, g\}, \{e, i\}), (\{a, d, f, j\}, \{c, d, h, g\}, \{e, i\}), (\{a, c, g, j\}, \{b, d, h, f, i\}, \{e\})$ and $(\{a, d, g, j\}, \{b, c, h, f, i\}, \{e\})$ are preferred labellings.

2.3.2.4 Stable Labelling and Semi-Stable Labelling

Definition 2.19 (Stable labelling and semi-stable labelling). Let $F = (A, R)$ be an argumentation framework and $\mathcal{L} : A \mapsto \{\text{IN, OUT, UNDEC}\}$ be a total function.

- $\mathcal{L}$ is a stable labelling, if and only if it is a complete labelling, and $undec(\mathcal{L}) = \emptyset$.
- $\mathcal{L}$ is a semi-stable labelling, if and only if it is a complete labelling, and $in(\mathcal{L}) \cup out(\mathcal{L})$ is maximal (with respect to set-inclusion).

Example 2.17. Consider again the argumentation framework in Example 2.11. According to Definition 2.19, it holds that

- There is no stable labelling;
- $(\{a, c, g, j\}, \{b, d, h, f, i\}, \{e\})$ and $(\{a, d, g, j\}, \{b, c, h, f, i\}, \{e\})$ are semi-stable labellings.

2.3.2.5 Ideal Labelling and Eager Labelling

The definition of ideal labelling is more complex. According to [9], in order to define the ideal labelling of an argumentation framework, we need to introduce the following notion [10].

Definition 2.20 (Bigger labelling). Let $\mathcal{L}_1$ and $\mathcal{L}_2$ be labellings of an argumentation framework $F = (A, R)$. It is said that $\mathcal{L}_2$ is bigger (also called *more or equally committed*) than $\mathcal{L}_1$ (written as $\mathcal{L}_1 \sqsubseteq \mathcal{L}_2$) if and only if $in(\mathcal{L}_1) \subseteq in(\mathcal{L}_2)$ and $out(\mathcal{L}_1) \subseteq out(\mathcal{L}_2)$.

The relation "$\sqsubseteq$" defines a partial order (reflective, anti-symmetric, transitive) on the labellings of an argumentation framework.

Definition 2.21 (Ideal labelling and eager labelling). Let $F = (A, R)$ be an argumentation framework. The ideal (eager) labelling is the biggest admissible labelling that is smaller than or equal to each preferred labelling (respectively, semi-stable labelling).

Example 2.18. Continue Examples 2.11 and 2.17.

- There is an admissible labelling $(\{a\}, \{b\}, \{c, d, e, f, g, h, i, j\})$ that is smaller than each preferred labelling. It is the ideal labelling of $F_{2.4}$.
- Meanwhile, there is an admissible labelling $(\{a, g\}, \{b, f, i\}, \{c, d, e, h, j\})$ that is smaller than each semi-stable labelling. It is the eager labelling of $F_{2.4}$.

2.3.3 Relations Between the Two Approaches

As introduced in [3], under most of argumentation semantics, there is a bijective correspondence between the set of labellings and sets of extensions of an argumentation framework. On the one hand, given an argumentation framework, the labels IN can be understood as identifying the members of an extension. According to [3], we have the following definition.

Definition 2.22 (A mapping from a labelling to an extension). Let $F = (A, R)$ be an argumentation framework, and $\mathfrak{L}$ be a labelling of F. The corresponding set of arguments $\text{Lab2Ext}(\mathfrak{L}) = in(\mathfrak{L})$.

On the other hand, given an extension E of an argumentation framework, if it is conflict-free, then we may construct a labelling such that the arguments belonging to E are labelled IN, those attacked by some arguments of E are labelled OUT, and those which neither belong to E nor are attacked by E are labelled UNDEC. Formally, we have the following definition.

Definition 2.23 (A mapping from an extension to a labelling). Let $F = (A, R)$ be an argumentation framework, and $E \subseteq A$ be a conflict-free set of arguments. The corresponding labelling, denoted as $\text{Ext2Lab}(E)$, is defined as $\text{Ext2Lab}(E) = (E, E^+, A \setminus (E \cup E^+))$, in which $E^+ = \{\alpha \in A \mid \exists \beta \in E, \text{ such that } (\beta, \alpha) \in R\}$.

As proved in [8], there is a correspondence between admissible sets and admissible labellings.

Proposition 2.1. *For any argumentation framework* $F = (A, R)$*, if* $\mathfrak{L}$ *is an admissible labelling of F then* $\text{Lab2Ext}(\mathfrak{L})$ *is an admissible set of F; if E is an admissible set of F then* $\text{Ext2Lab}(E)$ *is an admissible labelling of F.*

It should be noticed that the correspondence between admissible sets and admissible labellings is not bijective. This is because different admissible labellings may correspond to the same admissible set. For instance, with respect to $F_{2.2}$ in Example 2.4, $(\{a\}, \{b\}, \{c, d\})$ and $(\{a\}, \emptyset, \{b, c, d\})$ are two admissible labellings. Both of them give rise to the same admissible set $\{a\}$.

Except admissible semantics, under other semantics introduced above, there exists a bijective correspondence between sets of extensions and a set of labellings, of an argumentation framework.

In this book, we use σ to denote a semantics, which could be admissible (adm), complete (co), preferred (pr), grounded (gr), stable (st), semi-stable (sst), ideal (id) or eager (ea). Based on these notations, we have the following proposition.

Proposition 2.2. *Let* $F = (A, R)$ *be an argumentation framework, and* $\sigma \in \{co, pr, gr, st, sst, id, ea\}$ *be a semantics. For all* $\mathfrak{L} \in \mathcal{L}_\sigma(F)$*, there exists* $E \in \mathcal{E}_\sigma(F)$ *such that* $\mathfrak{L} = \text{Ext2Lab}(E)$*, and for all* $E \in \mathcal{E}_\sigma(F)$*, there exists* $\mathfrak{L} \in \mathcal{L}_\sigma(F)$ *such that* $E = \text{Lab2Ext}(\mathfrak{L})$*.*

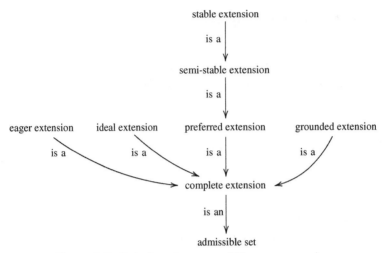

Figure 2.5   Relations between different semantics.

2.3.4 *Relations Between Different Semantics*

With respect to the extension-based approach, the relations between different argumentation semantics introduced above can be illustrated in Figure 2.5. According to Definitions 2.4 and 2.6, it holds that a complete extension is an admissible set. According to Definition 2.7, a grounded extension (respectively, a preferred extension) is a complete extension. Meanwhile, as proved in [10], an ideal extension (respectively, eager extension) is also a complete extension. Finally, according to [4], every stable extension is also a semi-stable extension, which in turn is a preferred extension. Given an argumentation framework, its eager extension is a superset of its ideal extension, which is in turn a superset of its grounded extension. So, the grounded semantics is the most skeptical semantics mentioned above. On the other hand, as to the labelling-based approach, we have similar results.

2.3.5 *Status of Arguments*

According to the extensions (labellings) of an argumentation framework, the status of arguments can be determined. Given a labelling, the status of each argument with respect to this labelling is IN, OUT, or UNDEC.

On the other hand, from the perspective of the extension-based approach, the status of arguments is determined by an extension, rather than a labeling. As presented in Definition 2.23, when an extension is conflict-free, it can be mapped to a labelling. Hence, given an argumentation framework and a conflict-free extension of it, for each argument within the argumentation framework, we may differentiate three status: *accepted*, *rejected* and *undecided*, corresponding to IN, OUT and UNDEC in the labelling-based approach.

Informally, an argument is *accepted* with respect to an extension, if and only if it belongs to this extension; an argument is *rejected* with respect to an extension, if and only if it is attacked by another argument that is accepted with respect to this extension; an argument is *undecided* with respect to an extension, if and only if it is neither accepted nor rejected with respect to this extension. Here, the extension should be conflict-free. Otherwise, the above three classes of status could not be differentiated. For instance, consider an argumentation framework $(\{a, b\}, \{(a, b)\})$. Suppose that $\{a, b\}$ is an extension. It follows that b is both accepted (because it belongs to $\{a, b\}$) and rejected (because it is attacked by a that is accepted with respect to $\{a, b\}$).

Definition 2.24 (Status of arguments with respect to an extension). Let $F = (A, R)$ be an argumentation framework, and E be an extension that is conflict-free. For all $\alpha \in A$:

- α is accepted with respect to E, if and only if $\alpha \in E$;
- α is rejected with respect to E, if and only if there exists $\beta \in E$, such that $(\beta, \alpha) \in R$;
- α is undecided with respect to E, if and only if α is neither accepted nor rejected with respect to E.

Beside the status of arguments with respect to an extension or a labelling, we may evaluate the status of arguments with respect to sets of extensions or a set of labellings, call the *justification status* of arguments. Since under some argumentation semantics, an argumentation framework might have multiple extensions (labellings), the justification status of an argument could be *sceptically justified*, *credulously justified*, and *indefensible*. An argument is sceptically justified, if it belongs to each extension (labelling); an argument is credulously justified, if it belongs to some (at least one) extensions (labellings) and does not belong to some other (at least one) extensions (labellings); an argument is indefensible, if it does not belong to any extension (labelling). Formally, we have the following definition.

Definition 2.25 (Justification status of arguments). Given $F = (A, R)$ and a conflict-free semantics σ, an argument $\alpha \in A$ is *credulously justified* if and only if $\exists E_1, E_2 \in \mathcal{E}_\sigma(F)$, such that $\alpha \in E_1$ and $\alpha \notin E_2$; it is *skeptically justified* if and only if $\forall E \in \mathcal{E}_\sigma(F), \alpha \in E$; and it is *indefensible* if and only if $\nexists E \in \mathcal{E}_\sigma(F)$ such that $\alpha \in E$.

2.4 Conclusions

In this chapter, we have introduced the notion of abstract argumentation frameworks and the semantics of an argumentation framework from the perspective of an extension-based approach and that of a labelling-based approach, respectively. Under most semantics (complete, preferred, grounded, stable, semi-stable, ideal and eager), there is a bijective correspondence between sets of extensions and a set of labellings.

Under a given semantics, an argumentation framework may have a unique extension or multiple extensions. If under a semantics σ, any argumentation framework has only one extension, then it is called a unique-status semantics; otherwise, it is called a multiple-status semantics. Grounded, ideal and eager are three unique-status semantics introduced in this chapter. Among them, the grounded semantics is the most skeptical one, the ideal semantics is the second, and the eager semantics is the third. On the other hand, admissible, complete, preferred, stable and semi-stable are multiple-status semantics. Under stable semantics, an argumentation framework might have an empty set of extension. Semi-stable semantics was proposed to handle this problem.

Based on the definition of argumentation semantics, given an argumentation framework, an important problem is to efficiently compute the status of arguments. In the subsequent chapters, we will focus on this problem.

References

[1] P.M. Dung, On the acceptability of arguments and its fundamental role in nonmonotonic reasoning, logic programming and n-person games, Artificial Intelligence 77 (2) (1995) 321–357.

[2] P. Baroni, M. Giacomin, On principle-based evaluation of extension-based argumentation semantics, Artificial Intelligence 171 (10–15) (2007) 675–700.

[3] P. Baroni, M. Caminada, M. Giacomin, An introduction to argumentation semantics, The Knowledge Engineering Review 26 (4) (2011) 365–410.

[4] M. Caminada, Semi-stable semantics, in: Proceedings of First International Conference on Computational Models of Argument, 2006, pp. 121–130.

[5] P.M. Dung, P. Mancarella, F. Toni, Computing ideal sceptical argumentation, Artificial Intelligence 171 (10–15) (2007) 642–674.

[6] H. Jakobovits, D. Vermeir, Robust semantics for argumentation frameworks, Journal of Logic and Computation 9 (2) (1999) 215–261.

[7] M. Caminada, On the issue of reinstatement in argumentation, in: Proceedings of the 10th European Conference on Logics in Artificial Intelligence, 2006, pp. 111–123.

[8] M. Caminada, Dov Gabbay, A logical account of formal argumentation, Studia Logica 93 (2–3) (2009) 109–145.

[9] M. Caminada, A labelling approach for ideal and stage semantics, Argument & Computation 2 (1) (2011) 1–21.

[10] M. Caminada, G. Pigozzi, On judgment aggregation in abstract argumentation, Journal of Autonomous Agents and Multi-Agent Systems 22 (1) (2011) 64–102.

CHAPTER 3

Existing Approaches for Computing Argumentation Semantics

Chapter Outline

3.1 Introduction

Given an argumentation framework, an important problem is to determine the status of arguments, i.e., to compute the extensions or labellings of the argumentation framework. In recent years, various approaches for computing the semantics of argumentation have been developed, including reduction approaches and direct approaches, as summarized in [1].

The basic idea of the reduction approaches is to exploit existing efficient methods that have been developed for other purposes. By translating the problems of finding extensions or labellings to the problems within other formalisms (such as propositional logic and answer-set programming, etc.), the sophisticated problem solvers dedicated to these formalisms could be exploited. On the other hand, the direct approaches (such as labelling-based algorithms, dialogue games, etc.) are developed from scratch.

In this chapter, we introduce a reduction approach that is based on answer-set programming (ASP), and a direct approach, called labelling-based algorithms.

Efficient Computation of Argumentation Semantics. http://dx.doi.org/10.1016/B978-0-12-410406-8.00003-8

3.2 Approaches Based on Answer Set Programming

Since many natural argumentation problems are in general intractable, one way to resolve them is to translate them to another language, for which sophisticated systems already exist. Answer set programming is a very good candidate for this purpose. This is because advanced solvers such as Smodels, DLV and Clasp which are able to deal with large problem instances are available. In other words, after we encode solutions to an argumentation problem into the intended models of a logic program, the existing ASP solvers can be exploited to compute some or multiple answer set(s) of the program together with the input, and the solutions of the problem can then be easily read off from the answer sets [2].

3.2.1 Answer Set Programming

Answer set programming (or briefly, ASP) is a declarative problem-solving paradigm, rooted in logic programming and non-monotonic reasoning [3]. Its syntax and semantics are introduced as follows.

3.2.1.1 Syntax

A term is a constant symbol or a variable symbol. An atom is an expression $p(t_1, \ldots, t_n)$, where p is a predicate of arity $n \geq 0$ and each t_i is a term. An atom is ground if it is free of variable. A literal is an atom $p(t_1, \ldots, t_n)$ or a strongly negated atom $\neg p(t_1, \ldots, t_n)$. Negation as failure is represented as *not*. A disjunctive rule r is a rule of the form:

$$l_1 \vee \ldots \vee l_k :- l_{k+1}, \ldots, l_m, not\ l_{m+1}, \ldots, not\ l_n \tag{3.1}$$

where $n \geq m \geq k \geq 0$, l_i $(1 \leq i \leq n)$ is a literal, and $\vee$ represents *epistemic disjunction*. The *head* of a rule r is the set $head(r) = \{l_0, \ldots, l_k\}$, and the body of r is $body(r) = \{l_{k+1}, \ldots, l_m, not\ l_{m+1}, \ldots, not\ l_n\}$. Furthermore, the *positive body* and *negative body* of r are denoted by $pos(r) = \{l_{k+1}, \ldots, l_m\}$ and $neg(r) = \{l_{m+1}, \ldots, l_n\}$, respectively.

A rule r is *normal* if $k \leq 1$ and a *constraint* if $k = 0$. A rule r is *safe* if each variable in r occurs in $pos(r)$. A rule r is ground if every atom in r is ground. A *fact* is a ground rule without disjunction and empty body. A program is a finite set of disjunctive rules. If each rule in a program is normal (respectively, ground), we call the program normal (respectively, ground).

3.2.1.2 Answer Set Semantics

Let Π be a ground disjunctive logic program and Lit_Π be the set of ground literals in the language of Π. First, let us consider the case that each rule in Π does not contain *not*. An answer set of Π is any minimal subset of $S \subseteq Lit_P$ such that,

(1) for each rule $l_1 \vee \ldots \vee l_k :- l_{k+1}, \ldots, l_m$ from Π, if $l_{k+1}, \ldots, l_m \in S$, then for some $i = 1, \ldots, k$, it holds that $l_i \in S$;

(2) if S contains a pair of complementary literals, then $S = Lit_\Pi$.

Second, when the rules in Π contain *not*, for any set $S \subseteq Lit_\Pi$, let Π^S be the logic program obtained from Π by deleting: (1) each rule that has a formula *not l* in its body with $l \in S$, and (2) all formulas of the form *not l* in the bodies of the remaining rules. After this treatment, Π^S does not contain *not*. Then, S is called an answer set of Π if and only if S is an answer set of Π^S.

3.2.2 ASP for Argumentation

As summarized by [4], several approaches have been proposed for computing extensions of argumentation frameworks by using ASP solvers. All of them rely upon the mapping of an argumentation framework into a logic program whose answer sets are in one-to-one correspondence with the extensions of the original argumentation framework. These approaches are classified into two groups: those which result in an argumentation-dependent logic program, and those which result in a logic program with an argumentation framework-dependent component and an argumentation framework-independent (meta-)logic program. In this book, we only introduce Egly at al's approach [2], which belongs to the second group. Readers may refer to [4] for a more comprehensive introduction.

In Egly et al's approach, an argumentation framework $F = (A, R)$ is mapped to a set of facts $\hat{F}$ to be included in logic programs defined for computing extensions under various semantics.

$$\hat{F} = \{\arg(\alpha) \mid \alpha \in A\} \cup \{\text{defeat}(\alpha, \beta) \mid (\alpha, \beta) \in R\}$$

For conflict-free sets, the logic program consists of $\hat{F}$ together with π_{cf}, which is defined as follows (here, we abuse the notation ";" as "," in a set):

$$\pi_{cf} = \{\text{in}(\alpha) : - \ \text{not out}(\alpha), \arg(\alpha);$$
$$\text{out}(\alpha) : - \ \text{not in}(\alpha), \arg(\alpha);$$
$$: - \ \text{in}(\alpha), \text{in}(\beta), \text{defeat}(\alpha, \beta)\}.$$

For stable extensions, two additional rules for stability test are required, and we have the following definition for π_{st}:

$$\pi_{st} = \pi_{cf} \cup \pi_{srules}$$
$$\pi_{srules} = \{\text{defeated}(\alpha) : - \ \text{in}(\beta), \text{defeat}(\beta, \alpha);$$
$$: - \ \text{out}(\alpha), \text{not defeated}(\alpha)\}.$$

The first rule of π_{srules} computes those arguments attacked by the current guess, while the constraint in π_{srules} eliminates those guesses where some argument not contained in the guess remains undefeated.

For admissible extensions, the logic program π_{adm} is defined as follows:

$$\pi_{adm} = \pi_{cf} \cup \{\text{defeated}(\alpha) : - \ in(\beta), \text{defeat}(\beta, \alpha);$$
$$: - \ in(\alpha), \text{defeat}(\beta, \alpha), \text{not defeated}(\beta)\}.$$

The first rule is the same as the one in π_{st}. The new constraint rules out sets containing a non-defended argument.

For complete extensions, the logic program π_{co} is defined as follows:

$$\pi_{co} = \pi_{adm} \cup \{\text{undefended}(\alpha) : - \ \text{defeat}(\beta, \alpha), \text{not defeated}(\beta);$$
$$: - \ out(\alpha), \text{not undefended}(\alpha)\}.$$

For grounded extensions, the logic program π_{gr} is obtained by mirroring the characteristic function presentation of this semantics. The program makes use of an arbitrary ordering $<$ over arguments (such an order is provided by all ASP-solvers). The program consists of three components. The first component $\pi_<$ uses the given ordering $<$ over arguments to derive corresponding predicates for infimum inf, supremum sup and successor $succ$:

$$\pi_< = \{\text{lt}(\alpha, \beta) : - \ \text{arg}(\alpha), \text{arg}(\beta), \alpha < \beta;$$
$$\text{nsucc}(\alpha, \gamma) : - \ \text{lt}(\alpha, \beta), \text{lt}(\beta, \gamma);$$
$$\text{succ}(\alpha, \beta) : - \ \text{lt}(\alpha, \beta), \text{not nsucc}(\alpha, \beta);$$
$$\text{ninf}(\beta) : - \ \text{lt}(\alpha, \beta);$$
$$\text{inf}(\alpha) : - \ \text{arg}(\alpha), \text{not ninf}(\alpha);$$
$$\text{nsup}(\alpha) : - \ \text{lt}(\alpha, \beta);$$
$$\text{sup}(\alpha) : - \ \text{arg}(\alpha), \text{not nsup}(\alpha)\}.$$

The second component computes all arguments defended (by all arguments currently IN) in the layers obtained using inf, sup and $succ$, as follows:

$$\pi_{defended} = \{\text{defended\_up\_to}(\alpha, \beta) : - \ \text{inf}(\beta), \text{arg}(\alpha), \text{not defeat}(\beta, \alpha);$$
$$\text{defended\_up\_to}(\alpha, \beta) : - \ \text{inf}(\beta), in(\gamma), \text{defeat}(\gamma, \beta), \text{defeat}(\beta, \alpha);$$
$$\text{defended\_up\_to}(\alpha, \beta) : - \ \text{succ}(\gamma, \beta), \ \text{defended\_up\_to}(\alpha, \gamma),$$
$$\text{not defeat}(\beta, \alpha);$$
$$\text{defended\_up\_to}(\alpha, \beta) : - \ \text{succ}(\gamma, \beta), \ \text{defended\_up\_to}(\alpha, \gamma), in(\xi),$$
$$\text{defeat}(\xi, \beta), \text{defeat}(\beta, \alpha);$$
$$\text{defended}(\alpha) : - \ \text{sup}(\beta), \text{defended\_up\_to}(\alpha, \beta)\}.$$

The third component simply imposes that all defended arguments should be IN:

$$\text{in}(\alpha) : - \text{ defended}(\alpha)$$

Hence, $\pi_{gr} = \pi_< \cup \pi_{defended} \cup \{\text{in}(\alpha) : - \text{ defended}(\alpha)\}$.

For preferred extensions, the so-called saturation technique is used: Having computed admissible extension S (characterized via predicates in$(\cdot)$ and out$(\cdot)$), we perform a second guess using new predicates, say inN$(\cdot)$ and outN$(\cdot)$, to represent a further guess $T \supset S$. In order to check whether the first guess (via in$(\cdot)$ and out$(\cdot)$) characterizes a preferred extension, we have to ensure that no guess of the second form (i.e., via inN$(\cdot)$ and outN$(\cdot)$) characterizes an admissible extension. The saturation model is defined as follows:

$$\pi_{satpref} = \{\text{inN}(\alpha) \vee \text{outN}(\alpha) : - \text{ out}(\alpha);$$
$$\text{inN}(\alpha) : - \text{ in}(\alpha);$$
$$\text{fail} : - \text{ eq};$$
$$\text{fail} : - \text{ inN}(\alpha), \text{inN}(\beta), \text{defeat}(\alpha, \beta);$$
$$\text{fail} : - \text{ inN}(\alpha), \text{outN}(\beta), \text{defeat}(\beta, \alpha), \text{undefeated}(\beta);$$
$$\text{inN}(\alpha) : - \text{ fail}, \text{arg}(\alpha);$$
$$\text{outN}(\alpha) : - \text{ fail}, \text{arg}(\alpha);$$
$$: - \text{ not fail}\}.$$

Then, we still need to define the rules for the predicates eq and undefeated$(\cdot)$, which are computed via predicates eq_upto$(\cdot)$ (respectively, undefeated_upto$(\cdot, \cdot)$) in the same manner as we used defended_upto$(\cdot)$ for defended$(\cdot)$ in the module $\pi_{defended}$.

$$\pi_{eq} = \{\text{eq\_upto}(\alpha) : - \text{ inf}(\alpha), \text{in}(\alpha), \text{inN}(\alpha);$$
$$\text{eq\_upto}(\alpha) : - \text{ inf}(\alpha), \text{out}(\alpha), \text{outN}(\alpha);$$
$$\text{eq\_upto}(\alpha) : - \text{ succ}(\beta, \alpha), \text{in}(\alpha), \text{inN}(\alpha), \text{eq\_upto}(\beta);$$
$$\text{eq\_upto}(\alpha) : - \text{ succ}(\beta, \alpha), \text{out}(\alpha), \text{outN}(\alpha), \text{eq\_upto}(\beta);$$
$$\text{eq} : - \text{ } sup(\alpha), \text{eq\_upto}(\alpha)\};$$
$$\pi_{undefeated} = \{\text{undefeated\_upto}(\alpha, \beta) : - \text{ inf}(\beta), \text{outN}(\alpha), \text{outN}(\beta);$$
$$\text{undefeated\_upto}(\alpha, \beta) : - \text{ inf}(\beta), \text{outN}(\alpha), \text{not defeat}(\beta, \alpha);$$
$$\text{undefeated\_upto}(\alpha, \beta) : - \text{ succ}(\gamma, \beta), \text{undefeated\_upto}(\alpha, \gamma),$$
$$\text{outN}(\beta);$$
$$\text{undefeated\_upto}(\alpha, \beta) : - \text{ succ}(\gamma, \beta), \text{undefeated\_upto}(\alpha, \gamma),$$
$$\text{not defeat}(\beta, \alpha);$$
$$\text{undefeated}(\alpha) : - \text{ } sup(\beta), \text{undefeated\_upto}(\alpha, \beta)\}.$$

Then, the logic program for preferred extensions is $\pi_{pr} = \pi_{adm} \cup \pi_< \cup \pi_{eq} \cup \pi_{undefeated} \cup \pi_{satpref}$.

It has been proved that the answer sets of $\pi_\sigma \cup \hat{F}$ ($\sigma \in \{cf, adm, co, gr, pr\}$) are in one-to-one correspondence with the extensions of the argumentation framework F under semantics σ [5].

3.3 Labelling-Based Algorithms

Labelling-based algorithms are a type of direct approach. They are based on the concept of argument labellings. In existing literature, various algorithms have been proposed for computing argument labellings. In recent years, Modgil and Caminada's labelling-based algorithms (or briefly, MC algorithms) [6] have received much attention and been compared with some newly proposed algorithms [7,8]. Now, let us introduce the MC algorithm for computing the grounded labelling and the preferred labellings of an argumentation framework.

3.3.1 The Computation of Grounded Labellings

As presented in [6], given an argumentation framework $F = (A, R)$, an algorithm for generating the grounded labelling of F starts by assigning IN to all arguments that are not attacked, and then iteratively: OUT is assigned to any argument that is attacked by an argument that has just been made IN, and then IN to those arguments all of whose attackers are OUT. Thus, the arguments assigned IN on each iteration, are those that are reinstated by the arguments assigned IN on the previous iteration. The iteration continues until no more new arguments are made IN or OUT. Any arguments that remain unlabelled are then assigned UNDEC.

Algorithm 3.1 Algorithm for grounded labelling

1: $\mathcal{L}_0 := (\emptyset, \emptyset, \emptyset)$
2: **repeat**
3:   $in(\mathcal{L}_{i+1}) = in(\mathcal{L}_i) \cup \{\alpha \mid \alpha$ is not labelled in $\mathcal{L}_i$, and $\forall \beta$: if $(\beta, \alpha) \in R$ then $\beta \in out(\mathcal{L}_i)\}$
4:   $out(\mathcal{L}_{i+1}) = out(\mathcal{L}_i) \cup \{\alpha \mid \alpha$ is not labelled in $\mathcal{L}_i$, and $\exists \beta$: $(\beta, \alpha) \in R$, and $\beta \in in(\mathcal{L}_{i+1})\}$
5: **until** $\mathcal{L}_{i+1} = \mathcal{L}_i$
6: **return** $\mathcal{L}_G = (in(\mathcal{L}_i), out(\mathcal{L}_i), A \setminus (in(\mathcal{L}_i) \cup out(\mathcal{L}_i)))$

3.3.2 The Computation of Preferred Labellings

The MC algorithm for computing preferred labellings is realized by computing admissible labellings that maximize the number of arguments that are legally IN.

3.3.2.1 Generating Admissible Labellings

The approach to generate an admissible labelling is to start with the all-in labelling (the labelling in which every argument is labelled IN). This labelling trivially satisfies the absence of arguments that are illegally OUT. In order to satisfy another requirement of an admissible labelling (without arguments that are illegally IN), we need a way of changing the label of an argument that is illegally IN, preferrably without creating any arguments that are illegally OUT. This is done using a sequence of *transition steps*.

A transition step basically takes an argument that is illegally IN and relabels it to OUT. It then checks if, as a result of this, one or more arguments have become illegally OUT. If this is the case, then these arguments are relabelled to UNDEC. More precisely, a transition step can be described as follows.

Definition 3.1 (Transition step). Let $\mathcal{L}$ be a labelling for $F = (A, R)$ and α be an argument that is illegally IN in $\mathcal{L}$. A transition step on α in $\mathcal{L}$ consists of the following:

- the label of α is changed from IN to OUT;
- for every $\beta \in \{\alpha\} \cup \{\gamma \mid (\alpha, \gamma) \in R\}$, if β is illegally OUT, then the label of β is changed from OUT to UNDEC.

It has been proved that each transition step preserves the absence of arguments that are illegally out [9]. In other words, during the transition, no arguments that are illegally OUT are generated.

Given an initial labelling $\mathcal{L}_0$, a sequence of successive transition steps applied on $\mathcal{L}_0$ is called a *transition sequence*, which is defined as follows [9].

Definition 3.2 (Transition sequence). A transition sequence is a list $[\mathcal{L}_0, \alpha_1, \mathcal{L}_1, \alpha_2, \mathcal{L}_2, \ldots, \alpha_n, \mathcal{L}_n](n \geq 0)$ where each $\alpha_i (1 \leq i \leq n)$ is an argument that is illegally IN in labelling $\mathcal{L}_{i-1}$ and every $\mathcal{L}_i$ is the result of doing a transition step of α_i on $\mathcal{L}_{i-1}$. A transition sequence is called terminated if and only if $\mathcal{L}_n$ does not contain any argument that is illegally IN.

Example 3.1. Consider Figure 2.3 in Example 2.7. Let $\mathcal{L}_0 = (\{a, b, c, d, e\}, \emptyset, \emptyset)$. According to Definition 2.14, b, c, d and e are illegally IN. There are several transition sequences. A possible sequence is $[\mathcal{L}_0, b, \mathcal{L}_1, c, \mathcal{L}_2, e, \mathcal{L}_3, d, \mathcal{L}_4]$, detailed as follows.

$$\mathcal{L}_0 = (\{a, b, c, d, e\}, \emptyset, \emptyset)$$
$$\downarrow \text{select } b$$
$$\mathcal{L}_1 = (\{a, c, d, e\}, \{b\}, \emptyset)$$
$$\downarrow \text{select } c$$
$$\mathcal{L}_2 = (\{a, d, e\}, \{b, c\}, \emptyset)$$
$$\downarrow \text{select } e$$

$$\mathcal{L}_3 = (\{a, d\}, \{b, e\}, \{c\})$$

$$\downarrow \text{ select } d$$

$$\mathcal{L}_4 = (\{a\}, \{b\}, \{c, d, e\})$$

It has been verified that for any finite argumentation framework, only a finite number of transition steps can be performed, and the final labelling is an admissible labelling.

3.3.2.2 Generating Preferred Labellings

The generation of preferred labellings is based on the generation of admissible labellings. Since a preferred labelling is an admissible labelling that maximizes the number of arguments that are legally IN, we need some mechanisms to compare and choose the labellings we have generated, in that some admissible labellings are not preferred labellings. First, consider the following example presented in [6].

Example 3.2.   Let $F_{3.1}$ be an argumentation framework (Figure 3.1). Given $\mathcal{L}_0 = (\{a, b, c\}, \varnothing, \varnothing)$, we have the following two possible transition steps. The first one is $[\mathcal{L}_0, b, \mathcal{L}_1]$, which is an admissible and complete labelling. The second one is $[\mathcal{L}_0, c, \mathcal{L}'_1, b, \mathcal{L}'_2]$, which is an admissible labelling, but not a complete labelling (see Figure 3.2).

According to the above example, when we have $\mathcal{L}_0 = (\{a, b, c\}, \varnothing, \varnothing)$, if we choose b rather than c, then we could avoid getting a non-complete labelling (that is not a preferred labelling), and improve the efficiency of computation. For this purpose, in terms of [6], we may choose an argument that is *super-illegally IN*, if such an argument is available.

Definition 3.3 (Super-illegally IN).   Let $\mathcal{L}$ be a labelling for $F = (A, R)$. An argument α in $\mathcal{L}$ that is illegally IN, is also super-illegally IN if and only if it is attacked by an argument β that is legally IN in $\mathcal{L}$, or UNDEC in $\mathcal{L}$.

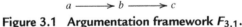

Figure 3.1   Argumentation framework $F_{3.1}$.

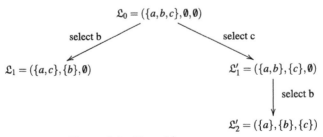

Figure 3.2   Transition sequences.

According to Definition 3.3, in Example 3.2, in $\mathfrak{L}_0 = (\{a, b, c\}, \emptyset, \emptyset)$, b is illegally IN, while c is not.

Based on the notion of super-illegally IN, an algorithm for preferred labellings is shown in Algorithm 3.2 [6].

The initial labelling $\mathfrak{L}_0 = (A, \emptyset, \emptyset)$, i.e., all arguments of the argumentation framework $F = (A, R)$ are labelled IN. With this initial labelling, the procedure *find_preferred_labellings*$(\mathfrak{L})$ iteratively applies transitions steps in an attempt to generate terminated

---

Algorithm 3.2 Algorithm for preferred labellings

1: *candidate_labellings* $:= \emptyset$
2: $\mathfrak{L}_0 := (A, \emptyset, \emptyset)$
3: *find_preferred_labellings*$(\mathfrak{L}_0)$
4: **return** *candidate_labellings*

5: **procedure** *find_preferred_labellings*$(\mathfrak{L})$
6: **if** $\exists \mathfrak{L}' \in candidate\_labellings : in(\mathfrak{L}) \subset in(\mathfrak{L}')$ **then return**
7: **if** $\mathfrak{L}$ does not have an argument that is illegally IN **then**
8:     **for** each $\mathfrak{L}' \in candidate\_labellings$ **do**
9:         **if** $in(\mathfrak{L}') \subset in(\mathfrak{L})$ **then**
10:            *candidate_labellings* $:=$ *candidate_labellings* $\setminus \{\mathfrak{L}'\}$;
11:         **end if**
12:     **end for**
13:     *candidate_labellings* $:=$ *candidate_labellings* $\cup \{\mathfrak{L}\}$;
14:     **return**;
15: **else**
16:     **if** $\mathfrak{L}$ has an argument that is super-illegally IN **then**
17:         $\alpha :=$ some argument that is super-illegally IN in $\mathfrak{L}$;
18:         *find_preferred_labellings*(*transition_step*$(\mathfrak{L}, \alpha)$);
19:     **else**
20:         **for** each α that is illegally IN is $\mathfrak{L}$ **do**
21:            *find_preferred_labellings*(*transition_step*$(\mathfrak{L}, \alpha)$)
22:         **end for**
23:     **end if**
24: **end if**
25: **end procedure**

---

transition sequences that update the global variable *candidate_labellings*. The algorithm preferentially selects from amongst super-illegally IN arguments for performing transition steps, if such arguments are available. If at any stage in the generation of a transition sequence, the arguments that are IN in the labelling $\mathfrak{L}_i$ thus far obtained are a strict subset of

$in(\mathcal{L}'_i)$ for some $\mathcal{L}'_i \in candidate\_labellings$, then no further transition steps on $\mathcal{L}_i$ can result in a preferred labelling (that maximizes the arguments that are IN). This follows from the result that during the course of a transition sequence, the set of IN labelled arguments monotonically decreases. Thus, any further transition steps on $\mathcal{L}_i$ will only reduce the arguments that are IN. In such cases, the algorithm backtracks to $\mathcal{L}_{i-1}$ and, if possible, selects another argument on which to perform a transition step. In the case that a transition sequence terminates, the obtained labelling $\mathcal{L}$ is compared with all labellings $\mathcal{L}'$ in $candidate\_$ $labellings$. If for any $\mathcal{L}'$, $in(\mathcal{L}')$ is a strict subset of $in(\mathcal{L})$, then $\mathcal{L}'$ is removed from $candidate\_labellings$. Thus, given a finite argumentation framework $F = (A, R)$, the algorithm calculates the preferred labellings and so preferred extensions.

3.4 Conclusions

In this chapter, we have introduced two approaches for computing argumentation semantics. Other approaches also exist. Among them, the methods based on dialogue games are famous. In a dialogue game, there are two players, a proponent and an opponent. The former argues in favour of an argument in question, and the latter argues against it. If the proponent has a winning strategy, then the argument is accepted. Some algorithms based on dialogue games can be found in [6] and [10].

References

[1] G. Charwat, W. Dvorak, S.A. Gaggl, J.P. Wallner, S. Woltran, Implementing Abstract Argumentation: A Survey, DBAI Technical Report, 2013.

[2] U. Egly, S.A. Gaggl, S. Woltran, ASPARTIX: implementing argumentation frameworks using answer-set programming, in: Proceedings of the 24th International Conference on Logic Programming, Elsevier, 2009, pp. 734–738.

[3] M. Gelfond, Answer Sets, Handbook of Knowledge Representation, 2007, pp. 285–316.

[4] F. Toni, M. Sergot, Argumentation and answer set programming, Lecture Notes in Computer Science Volume 6565 (2011) 164–180.

[5] U. Egly, S.A. Gaggl, S. Woltran, Answer-set programming encodings for argumentation frameworks, Argument & Computation 1 (2) (2010) 147–177.

[6] S. Modgil, M. CaminadaProof theories and algorithms for abstract argumentation frameworks, Argumentation in Artificial Intelligence, Springer, 2009.

[7] R. Baumann, G. Brewka, R. Wong, Splitting argumentation frameworks: an empirical evaluation, in: Proceedings of the 1st International Workshop on Theory and Applications of Formal Argumentation, 2012, pp. 17–31.

[8] S. Nofal, P. Dunne, K. Atkinson, On preferred extension enumeration in abstract argumentation, in: Proceedings of the 4th International Conference on Computational Models of Argument, 2012, pp. 205–216.

[9] M. Caminada, An algorithm for computing semi-stable semantics, in: Proceedings of the European Conference on Symbolic and Quantitative Approaches to Reasoning with Uncertainty, 2007, pp. 222–234.

[10] P.M. Thang, P.M. Dung, N.D. Hung, Towards a common framework for dialectical proof procedures in abstract argumentation, Journal of Logic and Computation 19 (6) (2009) 1071–1109.

Sub-Frameworks and Local Semantics

4.1 Introduction

In the previous two chapters, we only dealt with the semantics of a *whole* argumentation framework. In this book, we call it the *global semantics* of an argumentation framework. However, in many situations, it is better to focus on the semantics of a *part* of an argumentation framework, called the *local semantics* of the argumentation framework. For instance, when an argumentation framework changes by the addition (or the removal) of a set of arguments and/or a set of attacks, only the status of *affected* arguments is necessary to be determined [1]; when querying the status of some specific arguments, we may only pay attention to the set of *relevant* arguments [2]; and, for some argumentation frameworks, computing the status of arguments locally might dramatically reduce the computational complexity [3].

Given an argumentation framework and a subset of arguments within it, in order to define the local semantics of the argumentation framework with respect to this subset, we introduce a notion of *sub-frameworks*. Since the statuses of different arguments in an argumentation framework affect each other with respect to attack relation, a sub-framework might depend on (be conditioned by) some other sub-frameworks, which might in turn depend on some other sub-frameworks, and so on.

Efficient Computation of Argumentation Semantics. http://dx.doi.org/10.1016/B978-0-12-410406-8.00004-X

For a sub-framework which is conditioned by some other sub-frameworks, the definition of its semantics is not similar to the one for a Dung's argumentation framework (as presented in Section 2.3). Furthermore, from the perspective of computation, the approaches and algorithms for a Dung's argumentation framework should be modified to adapt to the characteristics of the sub-frameworks.

According to the above considerations, in this chapter, we focus on the following four basic issues:

- Notion of sub-frameworks;
- Dependence relation between deferent sub-frameworks;
- Semantics of sub-frameworks, which is regarded as the local semantics of a corresponding argumentation framework;
- Computation of the semantics of a sub-framework.

Related contents of this chapter and the subsequent chapters are originally presented in [1–3].

4.2 Notion of Sub-Frameworks

4.2.1 Informal Idea

Since an argumentation framework can be regarded as a directed graph, the notion of a sub-framework is similar to that of an *induced subgraph* of a directed graph. According to graph theory, given a graph $G = (E, V)$, a subgraph of G is a graph whose vertex set is a subset of that of G, and whose adjacency relation is a subset of that of G related to this subset. Furthermore, a subgraph H of a graph G is said to be induced if for any pair of vertices x and y of H, (x, y) is an edge of H if and only if (x, y) is an edge of G.

Example 4.1.   Let $F_{4.1}$ be an argumentation framework. The corresponding defeat graph is shown in Figure 4.1. Let us consider the following subgraphs.

- The subgraph induced by the subset $\{a, b\}$ is $(\{a, b\}, \{(a, b)\})$.
- The subgraph induced by the subset $\{c, d\}$ is $(\{c, d\}, \{(c, d)\})$.
- The subgraph induced by the subset $\{e, f\}$ is $(\{e, f\}, \{(e, f)\})$.

Now, an important problem arises: Can every subgraph in Example 4.1 be regarded as a sub-framework in which the status of arguments can be evaluated locally? First, the arguments in $(\{a, b\}, \{(a, b)\})$ or $(\{e, f\}, \{(e, f)\})$ are not attacked by any arguments outside the subgraph. It is intuitively feasible to evaluate the status of arguments locally. In other words,

$$a \longrightarrow b \longrightarrow c \longrightarrow d \qquad e \longrightarrow f$$

Figure 4.1   Argumentation framework $F_{4.1}$.

the status of arguments within the subgraph is not affected by the arguments outside the sub-framework, and therefore could be evaluated independently. Second, in $(\{c, d\}, \{(c, d)\})$, the argument c is attacked by the argument b, which is outside the subgraph. It is obvious that the status of the arguments c and d could not be evaluated locally in $(\{c, d\}, \{(c, d)\})$. In other words, when the arguments in an induced subgraph are attacked by some external arguments, we should take these external arguments into consideration. So, in this book, we will consider the following two classes of sub-frameworks.

If the arguments in a sub-framework are not attacked by any external arguments, then the sub-framework is called an *unconditioned sub-framework*. Otherwise, it is called a *conditioned sub-framework*. Hence, an unconditioned sub-framework is simply an induced subgraph of the corresponding defeat graph, while a conditioned sub-framework is composed of an induced subgraph and a *conditioning subgraph*. Here, the conditioning subgraph includes a set of nodes, each of which has at least a direct edge to a node of the induced subgraph. In Example 4.1, the conditioning sub-graph related to $(\{c, d\}, \{(c, d)\})$ is $(\{b\}, \{(b, c)\})$.

4.2.2 Formal Definition

Let $F = (A, R)$ be an argumentation framework and $B \subseteq A$ be a subset. According to Formula 2.1 in Definition 2.1, the set of outside parents of the arguments in B is B^-. In this book, we call B^- the set of *conditioning arguments* of B.

First, when $B^- = \emptyset$, the sub-framework induced by B is unconditioned. In this case, the sub-framework is represented as (B, R_B), in which $R_B = R \cap (B \times B)$.

Second, when $B^- \neq \emptyset$, the sub-framework induced by B is conditioned by the arguments in B^-. In this case, in terms of [1], the sub-framework induced by B is represented as $((B, R_B), (B^-, I_B))$, where (B^-, I_B) is a conditioning subgraph. Here, $I_B = R \cap (B^- \times B)$ is the set of interactions from the arguments in B^- to the arguments in B.

In this book, for simplicity, we regard (B, R_B) as a special case of $((B, R_B), (B^-, I_B))$ (when $B^- = \emptyset$ and $I_B = \emptyset$). So, all sub-frameworks are uniformly called *sub-frameworks*, and represented as $(B \cup B^-, R_B \cup I_B)$, in which B^- and I_B could be empty. Formally, we have the following definition.

Definition 4.1 (Sub-framework). Let $F = (A, R)$ be an argumentation framework, and $B \subseteq A$ be a set of arguments. Let $R_B = R \cap (B \times B)$ and $I_B = R \cap (B^- \times B)$. A sub-framework of F induced by B is a tuple:

$$(B \cup B^-, R_B \cup I_B) \tag{4.1}$$

For simplicity, when $B^- = \emptyset$ and $I_B = \emptyset$, $(B \cup B^-, R_B \cup I_B)$ is also denoted as (B, R_B).

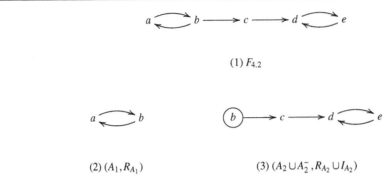

(1) $F_{4.2}$

(2) (A_1, R_{A_1}) (3) $(A_2 \cup A_2^-, R_{A_2} \cup I_{A_2})$

Figure 4.2 $F_{4.2}$ and its sub-frameworks. Argument b with a circle indicates that its status should be evaluated in the external sub-framework (A_1, R_{A_1}).

In addition, according to Definition 4.1, it is not required that the interactions from the arguments in B to the arguments in B^- should be empty. However, when they are not empty, although $(B \cup B^-, R_B \cup I_B)$ is still viewed as a sub-framework, there exists another sub-framework which is in turn conditioned by the arguments in B. As a result, the two sub-frameworks are dependent on each other. The notion of dependence relation between different sub-frameworks will be formulated in Definition 4.2.

Example 4.2.   Let $F_{4.2} = (A, R)$ be an argumentation framework (Figure 4.2(1)). Let $A_1 = \{a, b\}$ and $A_2 = \{c, d, e\}$. It follows that $R_{A_1} = \{(a, b), (b, a)\}$, $A_1^- = \emptyset$, $I_{A_1} = \emptyset$, $R_{A_2} = \{(c, d), (d, e), (e, d)\}$, $A_2^- = \{b\}$ and $I_{A_2} = \{(b, c)\}$. The sub-frameworks induced by A_1 and A_2 respectively are illustrated in Figures 4.2(2) and (3).

4.2.3 Dependence Relation Between Different Sub-Frameworks

Now, let us discuss the relations between different sub-frameworks. Given a sub-framework $(B \cup B^-, R_B \cup I_B)$, when $B^- \neq \emptyset$, it is restricted by some other sub-framework(s), which may be in turn restricted by some other sub-framework(s), and so on. Consider the following example:

Example 4.3.   The argumentation framework in Figure 4.3(1) could be decomposed into three sub-frameworks, illustrated in Figures 4.3(2), (3) and (4), indicated by SF_1, SF_2 and SF_3 respectively. We may observe that SF_1 is directly restricted by SF_3, and indirectly restricted by SF_2 and SF_1 itself.[1]

In this book, if a sub-framework SF is directly or indirectly restricted by another sub-framework SF', then we say that SF is dependent on SF'. Formally, we have the following definition.

---

[1] According to the example, when an argumentation framework contains cycles, different sub-frameworks in which some arguments belong to a cycle may restrict each other.

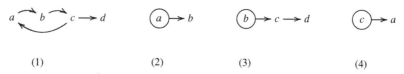

Figure 4.3 $F_{4.3}$ and its sub-frameworks.

Definition 4.2 (Dependence relation between sub-frameworks). Let $F = (A, R)$ be an argumentation framework, and $A_1, A_2 \subseteq A$ be subsets of A, $A_1 \neq A_2$. $(A_2 \cup A_2^-, R_{A_2} \cup I_{A_2})$ is *dependent* on $(A_1 \cup A_1^-, R_{A_1} \cup I_{A_1})$, if and only if $\exists \alpha \in A_1$ and $\beta \in A_2$ such that there is a path from α to β with respect to R. For convenience, $(A_1 \cup A_1^-, R_{A_1} \cup I_{A_1})$ is called a *conditioning sub-framework* of $(A_2 \cup A_2^-, R_{A_2} \cup I_{A_2})$.

Since a conditioned sub-framework $(B \cup B^-, R_B \cup I_B)$ may depend on some other sub-frameworks, before the status of arguments in B is evaluated, we hope that the status of arguments in B^- can be determined in advance. In this book, we only deal with a set of sub-frameworks over which there is a partial order.

4.3 Semantics of Sub-Frameworks

Given a sub-framework $(B \cup B^-, R_B \cup I_B)$ of an argumentation framework $F = (A, R)$, if $B^- = \emptyset$, then $(B \cup B^-, R_B \cup I_B)$ is an unconditioned sub-framework. In this case, $(B \cup B^-, R_B \cup I_B) = (B, R_B)$. In this book, the definitions of the semantics of (B, R_B) are the same as the ones presented in Section 2.3

On the other hand, when $B^- \neq \emptyset$, $(B \cup B^-, R_B \cup I_B)$ is a conditioned sub-framework. In this case, before the status of arguments in B is evaluated, the status of arguments in B^- should be determined in advance. Since $B^- \subseteq A$, there exists $C \subseteq A$, such that $B^- \subseteq C$ and $C^- = \emptyset$. Hence, it is possible that the status of arguments in B^- could be evaluated independently in an unconditioned sub-framework (C, R_C). Given a labelling or an extension of (C, R_C), the status of each argument in $B^- \subseteq C$ can be uniquely identified. According to the status of the arguments in B^-, the status of arguments in B is then evaluated.

Based on the above ideas, we introduce as follows the semantics of a conditioned sub-framework from the perspective of a labelling-based approach and of an extension-based approach, respectively.

4.3.1 Labellings of a Conditioned Sub-Framework

Let $(B \cup B^-, R_B \cup I_B)$ and (C, R_C) be sub-frameworks of $F = (A, R)$, and $B^- \subseteq C$. According to each labelling of (C, R_C), each argument in $B^- \subseteq C$ has a certain status (IN, OUT or UNDEC). After the status of arguments in B^- is labelled, the sub-framework

$(B \cup B^-, R_B \cup I_B)$ is called a *partially labelled sub-framework* (PLSF, for short). Formally, we have the following definition.

Definition 4.3 (Partially labelled sub-framework).   Let $(B \cup B^-, R_B \cup I_B)$ and (C, R_C) be sub-frameworks of $F = (A, R)$, such that $B^- \subseteq C$. Let $\mathcal{L}$ be a labelling of (C, R_C). We call $(B \cup B^-, R_B \cup I_B)^{\mathcal{L}}$ a partially labelled sub-framework, denoting that the labels of arguments in B^- conform to $\mathcal{L}$.

Given a partially labelled sub-framework $(B \cup B^-, R_B \cup I_B)^{\mathcal{L}}$, since the labels of the arguments in B^- conform to $\mathcal{L}$, we only need to assign new labels to the arguments in B. Formally, a labelling of a partially labelled sub-framework is defined as follows.

Definition 4.4 (Labelling of a partially labelled sub-framework).   Based on Definition 4.3, a *labelling* of $(B \cup B^-, R_B \cup I_B)^{\mathcal{L}}$ is defined as a total function

$$\mathcal{L}' : B \cup B^- \mapsto \{\text{IN, OUT, UNDEC}\}$$

such that for all $\alpha \in B^-$, $\mathcal{L}'(\alpha) = \mathcal{L}(\alpha)$.

According to Definition 4.4 , since the labels of arguments in B^- conform to $\mathcal{L}$, whether a label assigned to an argument in B is legal depends partially on $\mathcal{L}$.

Definition 4.5 (Legal/illegal labelling of a PLSF).   Based on Definition 4.3, let $\mathcal{L}'$ be a labelling of $(B \cup B^-, R_B \cup I_B)^{\mathcal{L}}$. For all $\alpha \in B$,

- α is legally IN in $\mathcal{L}'$ with respect to $\mathcal{L}$ if and only if α is labelled IN in $\mathcal{L}'$ and for all $\beta \in B$, if $(\beta, \alpha) \in R_B$ then β is labelled OUT in $\mathcal{L}'$; and for all $\beta \in B^-$, if $(\beta, \alpha) \in I_B$ then β is labelled OUT in $\mathcal{L}$ (and so in $\mathcal{L}'$);
- α is legally OUT in $\mathcal{L}'$ with respect to $\mathcal{L}$ if and only if α is labelled OUT in $\mathcal{L}'$ and there exists $\beta \in B$, such that $(\beta, \alpha) \in R_B$ and β is labelled IN in $\mathcal{L}'$; or there exists $\beta \in B^-$, such that $(\beta, \alpha) \in I_B$ and β is labelled IN in $\mathcal{L}$ (and so in $\mathcal{L}'$);
- α is legally UNDEC in $\mathcal{L}'$ with respect to $\mathcal{L}$ if and only if α is labelled UNDEC in $\mathcal{L}'$ and there exists no $\beta \in B \cup B^-$, such that $(\beta, \alpha) \in R_B \cup I_B$ and β is labelled IN in $\mathcal{L}$ or $\mathcal{L}'$, and it is not the case that: for all $\beta \in B \cup B^-$, if $(\beta, \alpha) \in R_B \cup I_B$ then β is labelled OUT in $\mathcal{L}$ or $\mathcal{L}'$;
- For $l \in \{$ IN, OUT, UNDEC$\}$, α is illegally l in $\mathcal{L}'$ with respect to $\mathcal{L}$ if and only if α is labelled l in $\mathcal{L}'$, but it is not legally l in $\mathcal{L}'$ with respect to $\mathcal{L}$.

Example 4.4.   Let us consider the two sub-frameworks in Example 4.2. Let $\mathcal{L}_1 = (\{a\}, \{b\}, \emptyset)$ and $\mathcal{L}_2 = (\emptyset, \{a, b\}, \emptyset)$ be two labellings of (A_1, R_{A_1}). So, for the sub-framework $(A_2 \cup A_2^-, R_{A_2} \cup I_{A_2})$, the corresponding two partially labelled sub-frameworks are $(A_2 \cup A_2^-, R_{A_2} \cup I_{A_2})^{\mathcal{L}_1}$ (in which Argument b is OUT according to $\mathcal{L}_1$) and $(A_2 \cup A_2^-, R_{A_2} \cup I_{A_2})^{\mathcal{L}_2}$ (in which Argument b is also OUT according to $\mathcal{L}_2$), as illustrated in Figure 4.4. Now, with respect to $(A_2 \cup A_2^-, R_{A_2} \cup I_{A_2})^{\mathcal{L}_1}$, let $\mathcal{L}_3 = (\{c, d\}, \{b, e\}, \emptyset)$. According to Definition 4.5, Argument c is legally IN in $\mathcal{L}_3$ with respect to $\mathcal{L}_1$ because it is

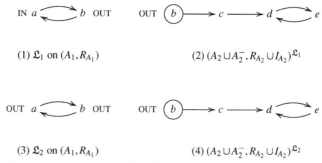

(1) $\mathcal{L}_1$ on (A_1, R_{A_1}) (2) $(A_2 \cup A_2^-, R_{A_2} \cup I_{A_2})^{\mathcal{L}_1}$

(3) $\mathcal{L}_2$ on (A_1, R_{A_1}) (4) $(A_2 \cup A_2^-, R_{A_2} \cup I_{A_2})^{\mathcal{L}_2}$

Figure 4.4 Two partially labelled sub-frameworks of $F_{4.2}$.

only attacked by Argument b, which is labelled OUT in $\mathcal{L}_1$ (and also in $\mathcal{L}_3$), while Argument d is illegally IN because it is attacked by Argument c which is labelled IN in $\mathcal{L}_3$.

Based on the notion of legal labelling of a partially labelled sub-framework, under admissible, complete, preferred, grounded, stable, semi-stable, ideal, and eager semantics, the labelling(s) of a partially labelled sub-framework can be defined as follows.

Definition 4.6 (Labelling-based semantics of a PLSF).   Based on Definition 4.3, let $\mathcal{L}'$ be a labelling of $(B \cup B^-, R_B \cup I_B)^{\mathcal{L}}$.

- $\mathcal{L}'$ is called an *admissible labelling* with respect to $\mathcal{L}$, if and only if the following two conditions hold:
    - $\mathcal{L}$ is an admissible labelling; and
    - each argument in B that is labelled IN in $\mathcal{L}'$ is legally IN in $\mathcal{L}'$ with respect to $\mathcal{L}$, and each argument in B that is labelled OUT in $\mathcal{L}'$ is legally OUT in $\mathcal{L}'$ with respect to $\mathcal{L}$.

- $\mathcal{L}'$ is called a *complete labelling* with respect to $\mathcal{L}$, if and only if the following two conditions hold:
    - $\mathcal{L}$ is a complete labelling; and
    - $\mathcal{L}'$ is an admissible labelling with respect to $\mathcal{L}$ and each argument in B that is labelled UNDEC in $\mathcal{L}'$ is legally UNDEC in $\mathcal{L}'$ with respect to $\mathcal{L}$.

- $\mathcal{L}'$ is called a *preferred labelling* with respect to $\mathcal{L}$, if and only if the following two conditions hold:
    - $\mathcal{L}$ is a preferred labelling; and
    - $\mathcal{L}'$ is a complete labelling with respect to $\mathcal{L}$, and $in(\mathcal{L}')$ is maximal (with respect to set-inclusion).

- $\mathcal{L}'$ is called a *grounded labelling* with respect to $\mathcal{L}$, if and only if the following two conditions hold:
    - $\mathcal{L}$ is a grounded labelling; and
    - $\mathcal{L}'$ is a complete labelling with respect to $\mathcal{L}$, and $in(\mathcal{L}')$ is minimal (with respect to set-inclusion).

- $\mathcal{L}'$ is called a *stable labelling* with respect to $\mathcal{L}$, if and only if the following two conditions hold:
  - $\mathcal{L}$ is a stable labelling; and
  - $\mathcal{L}'$ is a complete labelling with respect to $\mathcal{L}$, and $undec(\mathcal{L}') = \emptyset$.

- $\mathcal{L}'$ is called a *semi-stable labelling* with respect to $\mathcal{L}$, if and only if the following two conditions hold:
  - $\mathcal{L}$ is a semi-stable labelling; and
  - $\mathcal{L}'$ is a complete labelling with respect to $\mathcal{L}$, and $undec(\mathcal{L}')$ is minimal (with respect to set-inclusion).

- $\mathcal{L}'$ is called an *ideal labelling* with respect to $\mathcal{L}$, if and only if the following two conditions hold:
  - $\mathcal{L}$ is an ideal labelling; and
  - $\mathcal{L}'$ is the biggest admissible labelling with respect to $\mathcal{L}$ that is smaller than or equal to each preferred labelling with respect to $\mathcal{L}$.

- $\mathcal{L}'$ is called an *eager labelling* with respect to $\mathcal{L}$, if and only if the following two conditions hold:
  - $\mathcal{L}$ is an eager labelling; and
  - $\mathcal{L}'$ is the biggest admissible labelling with respect to $\mathcal{L}$ that is smaller than or equal to each semi-stable labelling with respect to $\mathcal{L}$.

According to Definition 4.6, the labelling $\mathcal{L}'$ is partially determined by the labelling $\mathcal{L}$. In Example 4.4, $\mathcal{L}_3$ is an admissible labelling with regard to $\mathcal{L}_1$, but it is not an admissible labelling with regard to $\mathcal{L}_2$. This is because $\mathcal{L}_1$ is an admissible labelling, while $\mathcal{L}_2$ is not.

4.3.2 Extensions of a Conditioned Sub-Framework

Let $(B \cup B^-, R_B \cup I_B)$ and (C, R_C) be sub-frameworks of an argumentation framework $F = (A, R)$, and $B^- \subseteq C$. From the perspective of the extension-based approach, the status of arguments in B^- is determined by an extension, rather than a labelling. According to the notion of the status (accepted, rejected or undecided) of arguments, we may assign a status to each argument in C. This process is called the *status assignment of arguments*.

Definition 4.7 (Status assignment of arguments). Let $F = (A, R)$ be an argumentation framework, and $C \subseteq A$ be a set of arguments. For all $E \in \mathscr{E}_\sigma(F)$, we may get a partition of C:

- $acc = \{\alpha \in C \mid \alpha$ is accepted with respect to $E\}$,
- $rej = \{\alpha \in C \mid \alpha$ is rejected with respect to $E\}$, and
- $undec = \{\alpha \in C \mid \alpha$ is undecided with respect to $E\}$.

We call $(acc, rej, undec)$ a status assignment of C with respect to E, denoted as $C[E] = (acc, rej, undec)$.

According to Definition 4.7, with respect to each extension E of (C, R_C), we have a status assignment of B^-, denoted as $B^-[E]$. It is obvious that if (C, R_C) has more than one extension, then for each extension of (C, R_C), there is a corresponding status assignment of B^-.

After the status of arguments in B^- is assigned with respect to E, the sub-framework $(B \cup B^-, R_B \cup I_B)$ is called a *partially assigned sub-framework* (PASF, for short), denoted as $(B \cup B^-, R_B \cup I_B)^E$.

Definition 4.8 (Partially assigned sub-framework). Let $(B \cup B^-, R_B \cup I_B)$ and (C, R_C) be sub-frameworks of an argumentation framework $F = (A, R)$, such that $B^- \subseteq C$. Let $E \subseteq C$ be a conflict-free set, $(B \cup B^-, R_B \cup I_B)^E$ is a partially assigned sub-framework, denoting that the status of arguments in B^- is assigned with respect to E.

Given a partially assigned sub-framework $(B \cup B^-, R_B \cup I_B)^E$, in which E is an extension of (C, R_C) under a semantics σ. Then, we say that a set $E' \subseteq B \cup B^-$ of arguments is an extension of $(B \cup B^-, R_B \cup I_B)^E$ with respect to E under the semantics σ if the following two conditions hold. First, the status of arguments in B^- is assigned with respect to E. Second, the status of argument in B is evaluated according to the criterion specified by σ. Formally, we have the following definition.

Definition 4.9 (Extension-based semantics of a PASF). Let (C, R_C) be an unconditioned sub-framework of an argumentation framework $F = (A, R)$, and $E \subseteq C$ be a conflict-free set of arguments. Let $(B \cup B^-, R_B \cup I_B)^E$ be a partially assigned sub-framework of F. Let $B^-[E] = (acc, rej, undec)$ be a status assignment of B^- with respect to E. Let $P \subseteq B$ be a set of arguments.

- P is *conflict-free* if and only if $\nexists \alpha, \beta \in P$, such that $(\alpha, \beta) \in R_B$.
- An argument $\alpha \in B$ is acceptable with respect to P and $B^-[E]$, if and only if the following two conditions hold:
  - $\forall \beta \in B$, if $(\beta, \alpha) \in R_B$, then $\exists \gamma \in P$, such that $(\gamma, \beta) \in R_B$, or $\exists \xi \in B^-$, such that $\xi \in acc$ (i.e., ξ is accepted with respect to E) and $(\xi, \beta) \in I_B$; and
  - $\forall \beta \in B^-$, if $(\beta, \alpha) \in I_B$, then $\beta \in rej$ (i.e., β is rejected with respect to E).
- $P \cup acc$ is admissible with respect to E if and only if the following two conditions hold:
  - E is admissible; and
  - P is conflict-free, and each argument in P is acceptable with respect to P and $B^-[E]$.
- $P \cup acc$ is a complete extension of $(B \cup B^-, R_B \cup I_B)^E$ with respect to E if and only if the following two conditions hold:

- E is a complete extension; and
- $P \cup acc$ is admissible with respect to E and each argument in B that is acceptable with respect to P and $B^-[E]$ is in P.

- $P \cup acc$ is a preferred extension of $(B \cup B^-, R_B \cup I_B)^E$ with respect to E if and only if the following two conditions hold:

  - E is a preferred extension; and
  - $P \cup acc$ is a maximal (with respect to set-inclusion) complete extension of $(B \cup B^-, R_B \cup I_B)^E$ with respect to E.

- $P \cup acc$ is a grounded extension of $(B \cup B^-, R_B \cup I_B)^E$ with respect to E if and only if the following two conditions hold:

  - E is a grounded extension; and
  - $P \cup acc$ is the minimal (with respect to set-inclusion) complete extension of $(B \cup B^-, R_B \cup I_B)^E$ with respect to E.

- $P \cup acc$ is a stable extension of $(B \cup B^-, R_B \cup I_B)^E$ with respect to E if and only if the following two conditions hold:

  - E is a stable extension; and
  - for all $\alpha \in B \setminus P$, α is attacked by $P \cup acc$.

- $P \cup acc$ is a semi-stable extension of $(B \cup B^-, R_B \cup I_B)^E$ with respect to E if and only if the following two conditions hold:

  - E is a semi-stable extension; and
  - $P \cup acc$ is a complete extension of $(B \cup B^-, R_B \cup I_B)^E$ with respect to E, such that $P \cup P^+$ is maximal (with respect to set-inclusion).

- $P \cup acc$ is ideal if and only if the following two conditions hold:

  - E is an ideal extension; and
  - $P \cup acc$ is the greatest (with respect to set-inclusion) admissible set with respect to E and it is contained in every preferred set of arguments of $(B \cup B^-, R_B \cup I_B)^E$ with respect to E.

- $P \cup acc$ is an eager extension if and only if the following two conditions hold:

  - E is an eager extension; and
  - $P \cup acc$ is the greatest (with respect to set-inclusion) admissible set (with respect to E) that is a subset of each semi-stable extension of $(B \cup B^-, R_B \cup I_B)^E$ with respect to E.

In this definition, it is clear that if $P \cup acc$ is admissible, then $P \cup acc$ is conflict-free. Otherwise, $\exists \alpha \in P$, such that α is attacked by an argument in acc. As a result, α is not acceptable with respect to P and $B^-[E]$. Contradiction.

(a) $(A_2 \cup A_2^-, R_{A_2} \cup I_{A_2})^{E_1}$ (b) $(A_2 \cup A_2^-, R_{A_2} \cup I_{A_2})^{E_2}$

Figure 4.5 Two partially assigned sub-frameworks of $F_{4.2}$.

Example 4.5. Let us consider again the sub-frameworks in Example 4.2. Under preferred semantics, (A_1, R_{A_1}) has two extensions $E_1 = \{a\}$ and $E_2 = \{b\}$. The status assignment of A_2^- with respect to E_1 and E_2 respectively is as follows:

$$A_2^-[E_1] = \{b\}[E_1] = (\emptyset, \{b\}, \emptyset)$$
$$A_2^-[E_2] = \{b\}[E_2] = (\{b\}, \emptyset, \emptyset)$$

Then, we get two partially assigned sub-frameworks $(A_2 \cup A_2^-, R_{A_2} \cup I_{A_2})^{E_1}$ and $(A_2 \cup A_2^-, R_{A_2} \cup I_{A_2})^{E_2}$ (Figure 4.5).

According to Definition 4.9, under preferred semantics, $(A_2 \cup A_2^-, R_{A_2} \cup I_{A_2})^{E_1}$ has one extension $E_3 = \{c, e\}$, while $(A_2 \cup A_2^-, R_{A_2} \cup I_{A_2})^{E_2}$ has two extensions $E_4 = \{b, d\}$ and $E_5 = \{b, e\}$. Here, the acceptability of arguments in each partially assigned sub-framework is related to the status of the arguments whose status is evaluated externally. For instance, in $(A_2 \cup A_2^-, R_{A_2} \cup I_{A_2})^{E_2}$, d is acceptable with respect to E_4 and $A_2^-[E_2] = (\{b\}, \emptyset, \emptyset)$ in that for the argument e in A_2 that attacks d, there exists d in E_4, such that $(d, e) \in R_{A_2}$, and for the argument c in A_2 that attacks d, there exists b in A_2^-, such that b is accepted with respect to E_2 and $(b, c) \in I_{A_2}$.

4.4 Computation of the Semantics of a Sub-Framework

As presented in the previous section, the semantics of a sub-framework could be formulated by the labelling-based approach or the extension-based approach. From the perspective of implementation, we may use the approaches mentioned in Chapter 3, with a slight modification.

In this section, we introduce two modified labelling-based algorithms for computing the preferred labellings and the grounded labelling of a partially labelled sub-framework, respectively.

As presented in Section 3.3, Modgil and Caminada's algorithms (or briefly, MC algorithms) are developed for a Dung's argumentation framework. For a partially labelled sub-framework $(B \cup B^-, R_B \cup I_B)^{\mathcal{L}}$, the MC algorithms should be modified such that the preferred labellings (or the grounded labelling) of a partially labelled sub-framework can be generated.

First, the algorithm for computing the preferred labelling of a partially labelled sub-framework is shown in Algorithm 4.1. Compared to the MC algorithm for computing preferred labellings, the modified algorithm is characteristic of the following three aspects.

First, since the labels of arguments in B^- should conform to $\mathfrak{L}$, the initial labelling of this algorithm is $(B \cup (in(\mathfrak{L}) \cap B^-), out(\mathfrak{L}) \cap B^-, undec(\mathfrak{L}) \cap B^-)$, in which all arguments in B are labelled IN, while the labels of arguments in B^- are assigned according to $\mathfrak{L}$.

Second, only the labels of arguments in B should be evaluated (to decide whether they are legal) and changed (by performing transition steps).

Third, when evaluating the legality of labels of arguments in B, the labelling $\mathfrak{L}$ should be taken into consideration. Due to this reason, the notions of *illegally IN*, *super-illegally IN* and *transition step* are different from the ones in the MC algorithm. Since the notion of *illegally IN* has been defined in Definition 4.5, we need only to present the other two notions as follows (Definitions 4.10 and 4.11).

Definition 4.10 (Super-illegally IN in a labelling of a PLSF). Let $(B \cup B^-, R_B \cup I_B)$ and (C, R_C) be sub-frameworks of an argumentation framework $F = (A, R)$, such that $B^- \subseteq C$. Let $\mathfrak{L}$ be a labelling of (C, R_C) under a semantics σ. Let $\mathfrak{L}'$ be a labelling of $(B \cup B^-, R_B \cup I_B)^{\mathfrak{L}}$. An argument $\alpha \in B$ that is illegally IN in $\mathfrak{L}'$ with respect to $\mathfrak{L}$, is also super-illegally IN in $\mathfrak{L}'$ with respect to $\mathfrak{L}$ if and only if:

- it is attacked by an argument $\beta \in B$ that is legally IN in $\mathfrak{L}'$ with respect to $\mathfrak{L}$ or UNDEC in $\mathfrak{L}'$ with respect to $\mathfrak{L}$; or
- it is attacked by an argument $\gamma \in B^-$ that is labelled IN or UNDEC in $\mathfrak{L}$.

Example 4.6. Continue Example 4.4. Let $\mathfrak{L}_4 = (\{b\}, \{a\}, \emptyset)$. With respect to $(A_2 \cup A_2^-, R_{A_2} \cup I_{A_2})^{\mathfrak{L}_4}$, let $\mathfrak{L}_5 = (\{b, c, d, e\}, \emptyset, \emptyset)$. According to Definition 4.10, Argument c is super-illegally IN in $\mathfrak{L}_5$ with respect to $\mathfrak{L}_4$, in that it is attacked by Argument b in A_2^- that is labelled IN in $\mathfrak{L}_4$.

Definition 4.11 (Transition step on an argument of a PLSF). Let $(B \cup B^-, R_B \cup I_B)$ and (C, R_C) be sub-frameworks of an argumentation framework $F = (A, R)$, such that $B^- \subseteq C$. Let $\mathfrak{L}$ be a labelling of (C, R_C) under a semantics σ. Let $\mathfrak{L}'$ be a labelling of $(B \cup B^-, R_B \cup I_B)^{\mathfrak{L}}$ and $\alpha \in B$ be an argument that is illegally IN in $\mathfrak{L}'$ with respect to $\mathfrak{L}$. A transition step on α in $\mathfrak{L}'$ consists of the following:

- the label of α is changed from IN to OUT;
- for every $\beta \in \{\alpha\} \cup \{\gamma \mid (\alpha, \gamma) \in R_B\}$, if β is illegally OUT in $\mathfrak{L}'$ with respect to $\mathfrak{L}$, then the label of β is changed from OUT to UNDEC.

Algorithm 4.1 Computing preferred labellings of a partially labelled sub-framework $(B \cup B^-, R_B \cup I_B)^{\mathfrak{L}}$

---

input : $\mathfrak{L} = (in(\mathfrak{L}), out(\mathfrak{L}), undec(\mathfrak{L})), sub = (B \cup B^-, R_B \cup I_B)$
output: *cand*

1: *cand* : $= \emptyset$ /* *Initially, the set of candidate labellings is empty.* */
2: $\mathfrak{L}' := (B \cup (in(\mathfrak{L}) \cap B^-), out(\mathfrak{L}) \cap B^-, undec(\mathfrak{L}) \cap B^-)$
3: *find_preferred_labellings*($\mathfrak{L}'$, *cand*, $B, \mathfrak{L}$)
4: **return** *cand*

5: **procedure** *find_preferred_labellings*($\mathfrak{L}'$, *cand*, $B, \mathfrak{L}$)
6: **if** $\exists \mathfrak{L}'' \in$ *cand* such that $in(\mathfrak{L}') \subset in(\mathfrak{L}'')$ **then return** /* *If* $in(\mathfrak{L}')$ *is a strict subset of* $in(\mathfrak{L}'')$, *then no further transition steps on* $\mathfrak{L}'$ *can result in a preferred labelling. Hence, the algorithm backtracks to select another argument for performing a transition step.* */
7: **if** there exists no argument in B that is illegally IN in $\mathfrak{L}'$ with respect to $\mathfrak{L}$ **then**
8:     **for** each $\mathfrak{L}'' \in$ *cand* **do**
9:         **if** $in(\mathfrak{L}'') \subset in(\mathfrak{L}')$ **then** *cand* $:= cand \setminus \{\mathfrak{L}''\}$; **end if**
        /* *If* $in(\mathfrak{L}'')$ *is a strict subset of* $in(\mathfrak{L}')$, *then* $\mathfrak{L}''$ *is removed from the set of candidate labellings.* */
10:     **end for**
11:     *cand* $:= cand \cup \{\mathfrak{L}'\}$;   /* *Add* $\mathfrak{L}'$ *as a new labelling.* */
12:     **return**;
13: **else**
14:     **if** there is an argument in B that is super-illegally IN in $\mathfrak{L}'$ with respect to $\mathfrak{L}$ **then**
15:         $\alpha :=$ some argument in B that is super-illegally IN in $\mathfrak{L}'$ with respect to $\mathfrak{L}$;
16:         *find_preferred_labellings*(*transition_step*($\mathfrak{L}', \alpha, \mathfrak{L}$), *cand*, $B, \mathfrak{L}$);
            /* *Since choosing different arguments that are super-illegally IN in* $\mathfrak{L}'$ *with respect to* $\mathfrak{L}$ *leads to the same result, we only need to try one of them.* */
17:     **else**
18:         **for each** α in B that is illegally IN in $\mathfrak{L}'$ with respect to $\mathfrak{L}$ **do**
19:             *find_preferred_labellings*(*transition_step*($\mathfrak{L}', \alpha, \mathfrak{L}$), *cand*, $B, \mathfrak{L}$)
20:         **end for**
21:     **end if**
22: **end if**
23: **end procedure**

Second, the algorithm for computing the grounded labelling of a partially labelled sub-framework is shown in Algorithm 4.2. Compared to the MC algorithm for computing the grounded labelling [4], there is only one difference, i.e., in Algorithm 4.2, the initial labelling is $\mathcal{L}_0 := (in(\mathcal{L}) \cap B^-, out(\mathcal{L}) \cap B^-, undec(\mathcal{L}) \cap B^-)$, rather than $\mathcal{L}_0 := (\emptyset, \emptyset, \emptyset)$. This means that the labels of arguments in B^- conform to $\mathcal{L}$.

Algorithm 4.2 Computing the grounded labelling of a partially labelled sub-framework $(B \cup B^-, R_B \cup I_B)^{\mathcal{L}}$

---

input : $sub = (B \cup B^-, R_B \cup I_B)$
$\qquad\qquad \mathcal{L} = (in(\mathcal{L}), out(\mathcal{L}), undec(\mathcal{L}))$

output: $\mathcal{L}_{GR}$

1: **procedure** *find_grounded_labellings(sub, $\mathcal{L}$)*
2: $\mathcal{L}_0 := (in(\mathcal{L}) \cap B^-, out(\mathcal{L}) \cap B^-, undec(\mathcal{L}) \cap B^-)$
$\quad i := -1$
3: **repeat**
4: $\quad i := i + 1$
5: $\quad in(\mathcal{L}_{i+1}) := in(\mathcal{L}_i) \cup \{\alpha \mid \alpha$ is not labelled in $\mathcal{L}_i$, and
$\qquad \forall \beta : $ if $(\beta, \alpha) \in R_B \cup I_B$ then $\beta \in out(\mathcal{L}_i)\}$
6: $\quad out(\mathcal{L}_{i+1}) := out(\mathcal{L}_i) \cup \{\alpha \mid \alpha$ is not labelled in $\mathcal{L}_i$, and
$\qquad \exists \beta : (\beta, \alpha) \in R_B \cup I_B$, and $\beta \in in(\mathcal{L}_{i+1})\}$
7: **until** $\mathcal{L}_{i+1} = \mathcal{L}_i$
8: **return** $\mathcal{L}_{GR} = (in(\mathcal{L}_i), out(\mathcal{L}_i), (B \cup B^-) \setminus (in(\mathcal{L}_i) \cup out(\mathcal{L}_i)))$
9: **end procedure**

4.5 Conclusions

In this chapter, we have defined two classes of sub-frameworks: unconditioned and conditioned. An unconditioned sub-framework is not dependent on any other sub-frameworks. Its semantics is the same as that of Dung's argumentation framework introduced in Chapter 2. On the other hand, a conditioned sub-framework is restricted by some external arguments. After the status of all external arguments has been determined, we get a partially labelled (respectively, assigned) sub-framework. The labellings and extensions of a partially labelled (respectively, assigned) sub-framework are defined subsequently. Finally, we have introduced in brief two algorithms for computing the preferred labellings and the grounded labelling of a partially labelled sub-framework, respectively.

The notion of sub-frameworks have also been introduced in [5,6]. Given an argumentation framework $F = (A, R)$ and a subset $S \subseteq A$, in [5], a sub-framework induced by S is called the restriction of F to S. However, the notion of a partially labelled (assigned) sub-framework has not been formally described in the existing literature. One important advantage of introducing the notion of a conditioned subframework is that given a subset of arguments that might be affected by some other arguments, their status could be computed locally.

References

[1] B. Liao, L. Jin, R.C. Koons, Dynamics of argumentation systems: a division-based method, Artificial Intelligence 175 (11) (2011) 1790–1814.

[2] B. Liao, H. Huang, Partial semantics of argumentation: basic properties and empirical results, Journal of Logic and Computation 23 (3) (2013) 541–562.

[3] B. Liao, Toward incremental computation of argumentation semantics: a decomposition-based approach, Annals of Mathematics and Artificial Intelligence 67 (3–4) (2013) 319–358.

[4] S. Modgil, M. Caminada, Proof theories and algorithms for abstract argumentation frameworks, Argumentation in Artificial Intelligence (2009) 105–129.

[5] P. Baroni, M. Giacomin, G. Guida, SCC-recursiveness: a general schema for argumentation semantics, Artificial Intelligence 168 (1–2) (2005) 162–210.

[6] R. Baumann, Splitting an argumentation framework, in: Proceedings of the 11th International Conference on Logic Programming and Nonmonotonic Reasoning, 2011, pp. 40–53.

CHAPTER 5

Relations between Global Semantics and Local Semantics

Chapter Outline

5.1 Introduction

In Chapters 2 and 4, we introduce the semantics of an argumentation framework and that of its sub-frameworks (called the *global semantics* and the *local semantics* of the argumentation framework, respectively). Now, an important question arises: what are the relations between these two semantics? More specifically, we have the following two sub-questions:

- Is there a mapping from the global semantics to the local semantics, of an argumentation framework? Is this mapping sound and complete?
- Is there a mapping from the local semantics to the global semantics, of an argumentation framework? Is this mapping sound and complete?

We will define in the subsequent sections these two kinds of mappings. The soundness and completeness of these mappings are affected by the types of argumentation semantics. Under some argumentation semantics such as admissible, complete, preferred and grounded, there exist sound and complete mappings. However, under some other argumentation semantics, sound and complete mappings might not exist. Let us consider the following example.

Efficient Computation of Argumentation Semantics. http://dx.doi.org/10.1016/B978-0-12-410406-8.00005-1

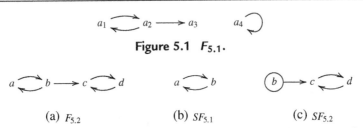

Figure 5.1 $F_{5.1}$.

(a) $F_{5.2}$ (b) $SF_{5.1}$ (c) $SF_{5.2}$

Figure 5.2 Argumentation framework $F_{5.2}$ and its sub-frameworks.

Example 5.1. Under stable semantics, the set of extensions of $F_{5.1}$ (as shown in Figure 5.1) is empty. Nevertheless, the sub-framework $(\{a_1, a_2, a_3\}, \{(a_1, a_2), (a_2, a_1), (a_2, a_3)\})$ has two stable extensions $\{a_1, a_3\}$ and $\{a_2\}$.

The above example shows that under stable semantics, there might not exist a mapping from global semantics to local semantics. In other words, the mapping from global semantics to local semantics might be incomplete. Meanwhile, there are also cases where the mapping from global semantics to local semantics is not sound. Let us consider the following example.

Example 5.2. Let $F_{5.2} = (\{a, b, c, d\}, \{(a, b), (b, a), (b, c), (c, d), (d, c)\})$, $SF_{5.1} = (\{a, b\}, \{(a, b), (b, a)\})$ and $SF_{5.2} = (\{b, c, d\}, \{(b, c), (c, d), (d, c)\})$, in which $SF_{5.1}$ is an unconditioned sub-framework, and $SF_{5.2}$ a conditioned sub-framework (see Figure 5.2). Under ideal semantics, $F_{5.2}$ has an extension $E_1 = \emptyset$, while $SF_{5.1}$ has an extension $E_2 = \emptyset$. In this case, for $SF_{5.2}$, there is only one partially assigned sub-framework: $SF_{5.2}^{E_2} = (\{b, c, d\}, \{(b, c), (c, d), (d, c)\})^{E_2}$, in which the status of Argument b is *undecided*. It follows that $SF_{5.2}^{E_2}$ has an ideal extension $E_3 = \{d\}$. So, if we define a mapping according to which the ideal extension of $SF_{5.2}^{E_2}$ is an empty set, then the mapping is not sound.

Given that under some argumentation semantics the mappings between global semantics and local semantics might be incomplete and/or unsound, in the subsequent sections and chapters, we only focus on several argumentation semantics (including admissible, complete, grounded and preferred) under which complete and sound mappings exist.

In addition, since there is a bijective correspondence between sets of extensions and a set of labellings under complete, grounded and preferred semantics, in this chapter, each mapping is formulated in terms of the extension-based approach or the labelling-based approach, rather than both approaches.

5.2 Mapping Global Semantics to Local Semantics

In terms of the extension-based approach, given an argumentation framework, a mapping from global semantics to local semantics is to restrict an extension of the argumentation framework

Figure 5.3 Argumentation framework $F_{5.3}$.

to those of its sub-frameworks. More specifically, for every extension of an argumentation framework, from the perspective of a sub-framework, only a part of the extension corresponds to the sub-framework. In other words, we may restrict the extension to the sub-framework (unconditioned or conditioned). Formally, we have the following definition.

Definition 5.1 (Restricting an extension to a sub-framework).   Let $F = (A, R)$ be an argumentation framework, and $(B \cup B^-, R_B \cup I_B)$ be a sub-framework of it. Let $\mathscr{E}_\sigma(F)$ be the set of extensions of F under a given semantics σ. When $\mathscr{E}_\sigma(F) \neq \emptyset$, for all $E \in \mathscr{E}_\sigma(F)$, the restriction of E to $(B \cup B^-, R_B \cup I_B)$ is defined as $E \cap (B \cup B^-)$.

Notice that when $B^- = \emptyset$, $(B \cup B^-, R_B \cup I_B) = (B, R_B)$ and $E \cap (B \cup B^-) = E \cap B$.

Example 5.3.   Consider an argumentation framework $F_{5.3}$ in Figure 5.3. Let $B = \{a_2, a_3, a_4\}$ and $C = \{a_1, a_2\}$. It follows that $B^- = \{a_1\}$, $R_B = \{(a_2, a_3), (a_3, a_4)\}$ and $I_B = \{(a_1, a_2)\}$, while $C^- = \emptyset$, $R_C = \{(a_1, a_2), (a_2, a_1)\}$ and $I_C = \emptyset$. Under complete semantics, $F_{5.3}$ has three extensions $E_1 = \{a_1, a_3, a_5\}$, $E_2 = \{a_2, a_4\}$ and $E_3 = \emptyset$. According to Definition 5.1, we have:

- $E_1 \cap C = \{a_1\}$ is the restriction of E_1 to (C, R_C);
- $E_2 \cap C = \{a_2\}$ is the restriction of E_2 to (C, R_C);
- $E_3 \cap C = \emptyset$ is the restriction of E_3 to (C, R_C);
- $E_1 \cap (B \cup B^-) = \{a_1, a_3\}$ is the restriction of E_1 to $(B \cup B^-, R_B \cup I_B)$;
- $E_2 \cap (B \cup B^-) = \{a_2, a_4\}$ is the restriction of E_2 to $(B \cup B^-, R_B \cup I_B)$;
- $E_3 \cap (B \cup B^-) = \emptyset$ is the restriction of E_3 to $(B \cup B^-, R_B \cup I_B)$.

Based on the notion of restricting an extension to a sub-framework, we may define the mapping from global semantics to local semantics.

First, for an unconditioned sub-framework, the mapping from global semantics to local semantics is closely related to the *directionality* of argumentation. The basic idea of directionality is that under some argumentation semantics, the status of an argument α is affected only by the status of its defeaters (which in turn are affected by their defeaters and so on), while the arguments which only receive an attack from α (and in turn those which are attacked by them and so on) do not have any effect on the status of α. So, with respect to an unconditioned sub-framework, under admissible, complete, preferred and grounded semantics which satisfy the directionality criterion, the mapping from global semantics to local semantics can be formulated by the following definition [1]:

Definition 5.2 (Mapping global semantics to local semantics with respect to an unconditioned sub-framework). Let $F = (A, R)$ be an argumentation framework, and (B, R_B) be an unconditioned sub-framework of F. Under a semantics $\sigma \in \{adm, co, pr, gr\}$,

$$\mathcal{E}_\sigma((B, R_B)) = \{E \cap B \mid E \in \mathcal{E}_\sigma(F)\} \tag{5.1}$$

Example 5.4. Continue Example 5.3. According to Formula 5.1, $E_1 \cap C = \{a_1\}$, $E_2 \cap C = \{a_2\}$ and $E_3 \cap C = \emptyset$ are complete extensions of (C, R_C).

Second, for a conditioned sub-framework $(B \cup B^-, R_B \cup I_B)$ of an argumentation framework $F = (A, R)$, for all $C \subseteq A$, if C is an unattacked set, and $B^- \subseteq C$, then (C, R_C) could be regarded as a conditioning sub-framework of $(B \cup B^-, R_B \cup I_B)$. Given $\sigma \in \{adm, co, pr, gr\}$, let $E \in \mathcal{E}_\sigma(F)$. For all $E_1 = E \cap C$, according to Formula 5.1, it holds that E_1 is an extension of (C, R_C) under semantics σ, and $(B \cup B^-, R_B \cup I_B)^{E_1}$ is a partially assigned sub-framework of F. Based on these notions, for a conditioned sub-framework $(B \cup B^-, R_B \cup I_B)$, the mapping from global semantics to local semantics can be formulated by the following proposition:

Proposition 5.1. *Let $F = (A, R)$ be an argumentation framework. Let B and C be subsets of A such that $B^- \neq \emptyset$, $C^- = \emptyset$ and $B^- \subseteq C$. Let $E_1 = E \cap C$. Under a semantics $\sigma \in \{adm, co, pr, gr\}$, $E \in \mathcal{E}_\sigma(F)$, if and only if $E \cap (B \cup B^-)$ is an extension of $(B \cup B^-, R_B \cup I_B)^{E_1}$.*

Since the soundness and completeness of the mapping from global semantics to local semantics for an unconditioned sub-framework (Formula 5.1) have been verified in [1], in this book, we only present the proofs of the soundness and completeness of the mapping from global semantics to local semantics for a conditioned sub-framework.

When proving Proposition 5.1, we found that the soundness of restricting an extension to a conditioned sub-framework under preferred and grounded semantics as well as the completeness of restricting an extension to a conditioned sub-framework under all four semantics depend on the property of the mappings from local semantics to global semantics. So, we present in this section the soundness of restricting an extension to a conditioned sub-framework under admissible and complete semantics, and then the two remaining parts at the end of Section 5.3.2.

Proof.   In this part, we prove the soundness of restricting an extension to a conditioned sub-framework under admissible and complete semantics, i.e., for all $\sigma \in \{adm, co\}$, if $E \in \mathcal{E}_\sigma(F)$, then $E \cap (B \cup B^-)$ is an extension of $(B \cup B^-, R_B \cup I_B)^{E_1}$.

- Under admissible semantics, E is an admissible set. Meanwhile, according to Formula 5.1, E_1 is admissible. Since E is admissible, it holds that E is conflict-free and each argument in E is acceptable with respect to E. In order to prove that $E \cap (B \cup B^-)$ is admissible with respect to E_1, since $E \cap B$ is obviously conflict-free, we only need to verify that each argument in $E \cap B$ is acceptable with respect to $E \cap B$ and $B^-[E_1]$, in which $B^-[E_1] = (acc, rej, undec)$ is a status assignment of B^- with respect to E_1, where $acc = \{\alpha \in B^- \mid \alpha$ is accepted with respect to $E_1\}, rej = \{\alpha \in B^- \mid \alpha$ is rejected with respect to $E_1\}$ and $undec = \{\alpha \in B^- \mid \alpha$ is undecided with respect to $E_1\}$.

Since each argument in E is acceptable with respect to E, for all $\alpha \in E \cap B, \alpha$ is acceptable with respect to E. Hence, for all $\beta \in A$, if $(\beta, \alpha) \in R$ then $\exists \gamma \in E$ such that $(\gamma, \beta) \in R$. Since $\alpha \in B$ and α is not attacked by the arguments in $A \setminus (B \cup B^-)$, it holds that $\beta \in B \cup B^-$. It follows that:

  - If $\beta \in B^-$, then: since β is attacked by the arguments in C, it holds that $\gamma \in E_1$, and therefore $\beta \in rej$ (satisfying the second condition of acceptability of arguments in a partially assigned sub-framework, in Definition 4.9).
  - Else, if $\beta \in B$, then: since β is attacked by the arguments in B or B^-, it holds that $\gamma \in E \cap B$ or $\gamma \in E \cap B^- = E_1 \cup B^- = acc$ (satisfying the first condition of acceptability of arguments in a partially assigned sub-framework, in Definition 4.9).

As a result, α is acceptable with respect to $E \cap B$ and $B^-[E_1]$.
- Under complete semantics, E is a complete extension. Based on the proof of the previous item, it holds that $E \cap (B \cup B^-)$ is admissible with respect to E_1. We need only to prove that each argument α in B that is acceptable with respect to $E \cap B$ and $B^-[E_1]$ is in $E \cap B$. Since α is only possibly attacked by the arguments in B or B^-, when α is acceptable with respect to $E \cap B$ and $B^-[E_1]$, we have the following two cases:

  - If α is attacked by $\beta \in B$, then according to the first condition of acceptability of arguments in a partially assigned sub-framework (Definition 4.9), $\exists \gamma \in E \cap B \subseteq E$, such that $(\gamma, \beta) \in R_B \subseteq R$, or $\exists \xi \in B^-$, such that $\xi \in acc = E_1 \cap B^- \subseteq E$ and $(\xi, \beta) \in I_B \subseteq R$; and
  - If α is attacked by $\beta \in B^-$, then according to the second condition of acceptability of arguments in a partially assigned sub-framework (Definition 4.9), β is rejected with respect to E_1, i.e., $\exists \gamma \in E_1 \subseteq E$, such that $(\gamma, \beta) \in R_C \subseteq R$.

As a result, for any argument β that attacks α, there exists an argument in E that attacks β. Therefore, α is acceptable with respect to E. Since every argument in $A \supseteq B$ that is acceptable with respect to E is in E, it holds that α is in E. Since $\alpha \in B$, we have $\alpha \in B \cap C$ or $\alpha \in B \setminus C$. If $\alpha \in B \cap C$, then $\alpha \in E \cap B$. Else, if $\alpha \in B \setminus C$, then since $\alpha \notin E_1$ (in that $\alpha \notin C$), it holds that α is in $E \cap B$ (otherwise, $\alpha \notin (E \cap B) \cup E_1 = E$, contradicting $\alpha \in E$). $\qquad \square$

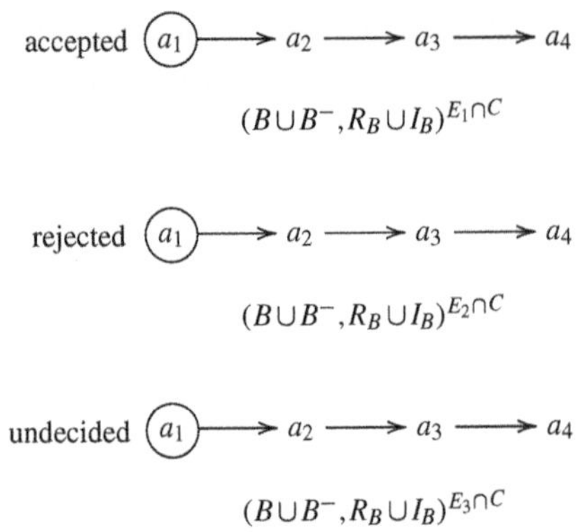

accepted $(a_1) \longrightarrow a_2 \longrightarrow a_3 \longrightarrow a_4$

$$(B \cup B^-, R_B \cup I_B)^{E_1 \cap C}$$

rejected $(a_1) \longrightarrow a_2 \longrightarrow a_3 \longrightarrow a_4$

$$(B \cup B^-, R_B \cup I_B)^{E_2 \cap C}$$

undecided $(a_1) \longrightarrow a_2 \longrightarrow a_3 \longrightarrow a_4$

$$(B \cup B^-, R_B \cup I_B)^{E_3 \cap C}$$

Figure 5.4   Three partially assigned sub-frameworks of $F_{5.3}$.

Example 5.5.   Continue Examples 5.3 and 5.4. Since $B^- \subseteq C$, (C, R_C) is a conditioning sub-framework of $(B \cup B^-, R_B \cup I_B)$. According to three complete extensions of (C, R_C), there are three partially assigned sub-frameworks of $(B \cup B^-, R_B \cup I_B)$ (as shown in Figure 5.4):

- $(B \cup B^-, R_B \cup I_B)^{E_1 \cap C}$,
- $(B \cup B^-, R_B \cup I_B)^{E_2 \cap C}$, and
- $(B \cup B^-, R_B \cup I_B)^{E_3 \cap C}$.

Then, according to Proposition 5.1 and the above proof, it holds that:

- $E_1 \cap (B \cup B^-) = \{a_1, a_3\}$ is a complete extension of $(B \cup B^-, R_B \cup I_B)^{E_1 \cap C}$,
- $E_2 \cap (B \cup B^-) = \{a_2, a_4\}$ is a complete extension of $(B \cup B^-, R_B \cup I_B)^{E_2 \cap C}$, and
- $E_3 \cap (B \cup B^-) = \emptyset$ is a complete extension of $(B \cup B^-, R_B \cup I_B)^{E_3 \cap C}$.

5.3 Mapping Local Semantics to Global Semantics

Given an argumentation framework, a mapping from local semantics to global semantics is to combine sets of extensions (labellings) of a set of sub-frameworks to form a set of extensions (respectively, labellings) of the argumentation framework. Since there are two kinds of sub-frameworks (conditioned and unconditioned) and various dependence relations between different sub-frameworks, there are many types of combinations of sub-frameworks, such as a combination of two unconditioned sub-frameworks, a combination of two conditioned sub-frameworks without dependence relation from one to another, a combination of an unconditioned sub-framework and a conditioned sub-framework in which the latter is fully conditioned by the former, and a combination of an unconditioned sub-framework and a conditioned sub-framework in which the latter is partially conditioned by the former, etc. In

Figure 5.5 Sub-frameworks of $F_{5.4}$.

this book, we study the first three combinations, in terms of the extension-based approach or the labelling-based approach.

5.3.1 Combining Extensions of Two Unconditioned Sub-Frameworks

Let (B_1, R_{B_1}) and (B_2, R_{B_2}) be two unconditioned sub-frameworks of an argumentation framework $F = (A, R)$. The (syntactic) combination of them, denoted as $(B_1 \cup B_2, R_{B_1} \cup R_{B_2})$, is also an unconditioned sub-framework.

Example 5.6. Let $F_{5.4} = (A, R)$ be an argumentation framework (Figure 5.5). Let $B_1 = \{a_1, a_2, a_7\}$, $B_2 = \{a_5, a_6\}$ and $B_3 = \{a_1, a_2, a_3, a_4\}$. Since $B_1^- = \emptyset$, $B_2^- = \emptyset$ and $B_3^- = \emptyset$, (B_1, R_{B_1}), (B_2, R_{B_2}) and (B_3, R_{B_3}) are unconditioned sub-frameworks. Let us consider the following two combined sub-frameworks.

- $(B_1 \cup B_2, R_{B_1} \cup R_{B_2})$, in which $B_1 \cap B_2 = \emptyset$.
- $(B_1 \cup B_3, R_{B_1} \cup R_{B_3})$, in which $B_1 \cap B_3 \neq \emptyset$.

The former is a combination of two unconditioned sub-frameworks that have no arguments in common, while the latter is a combination of two unconditioned sub-frameworks that have some arguments in common. In this book, the former is regarded as a special case of the latter.

Let $Int = B_1 \cap B_2$ denote the *intersection* of B_1 and B_2. Under a semantics σ under which every argumentation framework has at least one extension, the notion of combined extensions is defined as follows.

Definition 5.3 (Combined extensions of two unconditioned sub-frameworks). Let (B_1, R_{B_1}) and (B_2, R_{B_2}) be two unconditioned sub-frameworks of an argumentation framework $F = (A, R)$, and $Int = B_1 \cap B_2$. Let σ be a semantics under which every argumentation framework has at least one extension. The set of combined extensions of $(B_1 \cup B_2, R_{B_1} \cup R_{B_2})$, denoted as $CombExt_\sigma((B_1 \cup B_2, R_{B_1} \cup R_{B_2}))$, is defined as:

$$CombExt_\sigma((B_1 \cup B_2, R_{B_1} \cup R_{B_2})) = \{E_1 \cup E_2 \mid$$
$$E_1 \in \mathscr{E}_\sigma((B_1, R_{B_1})) \wedge E_2 \in \mathscr{E}_\sigma((B_2, R_{B_2})) \wedge (E_1 \cap Int = E_2 \cap Int)\}$$

Proposition 5.2. *Let (B_1, R_{B_1}) and (B_2, R_{B_2}) be two unconditioned sub-frameworks of an argumentation framework $F = (A, R)$, and $Int = B_1 \cap B_2$. For each $\sigma \in \{adm, co, gr, pr\}$, it holds that: $CombExt_\sigma((B_1 \cup B_2, R_{B_1} \cup R_{B_2})) = \mathscr{E}_\sigma((B_1 \cup B_2, R_{B_1} \cup R_{B_2}))$.*

Proof. $\Rightarrow$(Soundness): For all $E \in CombExt_\sigma((B_1 \cup B_2, R_{B_1} \cup R_{B_2}))$, it holds that $E \in \mathcal{E}_\sigma((B_1 \cup B_2, R_{B_1} \cup R_{B_2}))$.

(1) Under admissible semantics, for all $E_1 \in \mathcal{E}_{adm}((B_1, R_{B_1}))$ and $E_2 \in \mathcal{E}_{adm}((B_2, R_{B_2}))$, if $E_1 \cap Int = E_2 \cap Int$ then $E_1 \cup E_2 \in \mathcal{E}_{adm}((B_1 \cup B_2, R_{B_1} \cup R_{B_2}))$. This is because:

  – $E_1 \cup E_2$ is conflict-free. Assume the contrary. Then, $\exists \alpha \in E_1$ and $\beta \in E_2$, such that $(\alpha, \beta) \in R$. Since $\alpha \in E_1$ and $\beta \in E_2$, it holds that $\alpha \in B_1$ and $\beta \in B_2$. Since $(\alpha, \beta) \in R$, $\beta \in B_2$, and (B_2, R_{B_2}) is an unconditioned sub-framework, it holds that $\alpha \in B_2$. Since $\alpha \in E_1, \alpha \in B_1$ and $\alpha \in B_2$, it holds that $\alpha \in E_1 \cap B_1 \cap B_2 = E_1 \cap Int$. Since $E_1 \cap Int = E_2 \cap Int$, it follows that $\alpha \in E_2 \cap Int$, and therefore $\alpha \in E_2$. As a result, E_2 is not conflict-free. Contradiction.

  – $\forall \alpha \in E_1 \cup E_2$, α is acceptable with respect to $E_1 \cup E_2$. Since $\alpha \in E_i$ ($i = 1, 2$), then since $E_i \subseteq B_i$ and $B_i^- = \varnothing$, α is only attacked by the arguments in B_i. Since α is acceptable with respect to E_i, $\forall \beta \in B_i \subseteq B_1 \cup B_2$, if $(\beta, \alpha) \in R$, then $\exists \gamma \in E_1 \subseteq E_1 \cup E_2$, such that $(\gamma, \beta) \in R$. In other words, α is acceptable with respect to $E_1 \cup E_2$.

(2) Under complete semantics, we need to prove that $\forall \alpha \in B_1 \cup B_2$, if α is acceptable with respect to $E_1 \cup E_2$, then $\alpha \in E_1 \cup E_2$. For $\alpha \in B_i$ ($i = 1, 2$), since α is acceptable with respect to $E_1 \cup E_2$ and only affected by the arguments in B_i, α is acceptable with respect to $(E_1 \cup E_2) \cap B_i = E_i$; since every argument that is acceptable with respect to E_i is in E_i, it holds that α is in $E_i \subseteq E_1 \cup E_2$.

(3) Under preferred semantics, for all $E_1 \in \mathcal{E}_{pr}(B_1, R_{B_1})$ and $E_2 \in \mathcal{E}_{pr}(B_2, R_{B_2})$, since a preferred extension is also a complete extension, based on the proof in the case of complete semantics, we only need to prove that $E_1 \cup E_2$ is a maximal (with respect to set-inclusion) complete extension of $(B_1 \cup B_2, R_{B_1} \cup R_{B_2})$. Assume the contrary. Then, $\exists E' \subseteq B_1 \cup B_2$, such that E' is a preferred extension of $(B_1 \cup B_2, R_{B_1} \cup R_{B_2})$ and $E' \supset E_1 \cup E_2$. Let $\Phi = E' \setminus (E_1 \cup E_2)$. It holds that $\Phi \neq \varnothing$. Let $\Phi_1 = \Phi \cap B_1$ and $\Phi_2 = \Phi \cap B_2$. Since $\Phi \subseteq B_1 \cup B_2$, it holds that $\Phi_1 \cup \Phi_2 = \Phi \neq \varnothing$. There are the following two possible cases.

  – When $\Phi_1 \neq \varnothing$, since $E_1 \subseteq B_1$, $E_2 \cap B_1 = (E_2 \cap B_2) \cap B_1 = E_2 \cap Int$, and $E_2 \cap Int = E_1 \cap Int$, it holds that $(E_1 \cup E_2) \cap B_1 = (E_1 \cap B_1) \cup (E_2 \cap B_2) = (E_1 \cap B_1) \cup (E_2 \cap Int) = (E_1 \cap B_1) \cup (E_1 \cap Int) = E_1 \cap (B_1 \cup Int) = E_1 \cap B_1 = E_1$. So, $E_1 \cup \Phi_1 = ((E_1 \cup E_2) \cap B_1) \cup (\Phi \cap B_1) = ((E_1 \cup E_2) \cup \Phi) \cap B_1 = E' \cap B_1$. According to Formula 5.1, it holds that $E' \cap B_1$ is a complete extension of $(B_1 \cup B_2, R_{B_1} \cup R_{B_2})$. As a result, E_1 is not a preferred extension, contradicting $E_1 \in \mathcal{E}_{pr}((B_1, R_{B_1}))$.

  – When $\Phi_2 \neq \varnothing$, similarly, it turns out that E_2 is not a preferred extension, contradicting $E_2 \in \mathcal{E}_{pr}((B_2, R_{B_2}))$.

  As a result, $E_1 \cup E_2$ is a preferred extension of $(B_1 \cup B_2, R_{B_1} \cup R_{B_2})$.

(4) Under grounded semantics, similar to the proof in the case of preferred semantics, we may verify that $E_1 \cup E_2$ is the minimal (with respect to set-inclusion) complete extension of $(B_1 \cup B_2, R_{B_1} \cup R_{B_2})$.

$\Rightarrow$(Completeness): For all $E \in \mathcal{E}_\sigma((B_1 \cup B_2, R_{B_1} \cup R_{B_2}))$, $E \in CombExt_\sigma((B_1 \cup B_2, R_{B_1} \cup R_{B_2}))$.

According to Formula 5.1, $E_1 = E \cap B_1 \in \mathcal{E}_\sigma((B_1, R_{B_1}))$ and $E_2 = E \cap B_2 \in \mathcal{E}_\sigma((B_2, R_{B_2}))$. Meanwhile, since $E_1 \cap Int = (E \cap B_1) \cap Int = E \cap Int$ and $E_2 \cap Int = (E \cap B_2) \cap Int = E \cap Int$, it holds that $E_1 \cap Int = E_2 \cap Int$. According to Definition 5.3, $E \in CombExt_\sigma((B_1 \cup B_2, R_{B_1} \cup R_{B_2}))$.

Example 5.7. Continue Example 5.6. Let us consider the sub-frameworks (B_1, R_{B_1}) and (B_3, R_{B_3}). Under preferred semantics, the extensions of (B_1, R_{B_1}) are $E_1 = \{a_1, a_7\}$ and $E_2 = \{a_2\}$, while the extensions of (B_1, R_{B_1}) are $E_3 = \{a_1, a_3\}$ and $E_4 = \{a_2, a_4\}$. Let $Int = B_1 \cap B_3 = \{a_1, a_2\}$. According to Definition 5.3 and Proposition 5.2, we may combine the extensions of (B_1, R_{B_1}) and those of (B_3, R_{B_3}) to obtain the extensions of $(B_1 \cup B_3, R_{B_1} \cup R_{B_3})$.

- $E_1 \cup E_3 = \{a_1, a_3, a_7\}$ is a preferred extension of $(B_1 \cup B_3, R_{B_1} \cup R_{B_3})$, in that $E_1 \in \mathcal{E}_{pr}((B_1, R_{B_1}))$, $E_3 \in \mathcal{E}_{pr}((B_3, R_{B_3}))$, and $E_1 \cap Int = E_3 \cap Int = \{a_1\}$.
- $E_2 \cup E_4 = \{a_2, a_4\}$ is a preferred extension of $(B_1 \cup B_3, R_{B_1} \cup R_{B_3})$, in that $E_2 \in \mathcal{E}_{pr}((B_1, R_{B_1}))$, $E_3 \in \mathcal{E}_{pr}((B_3, R_{B_3}))$, and $E_2 \cap Int = E_4 \cap Int = \{a_2\}$.

5.3.2 Combining Extensions of a Conditioned Sub-Framework and Those of an Unconditioned Sub-Framework

Let $F = (A, R)$ be an argumentation framework. Let $(B \cup B^-, R_B \cup I_B)$ be a conditioned sub-framework, and (C, R_C) be an unconditioned sub-framework, of F. When combining $(B \cup B^-, R_B \cup I_B)$ and (C, R_C), we only consider the case when $B^- \subseteq C$. In this case, we have the following proposition.

Proposition 5.3. *Let $(B \cup B^-, R_B \cup I_B)$ be a conditioned sub-framework and (C, R_C) be an unconditioned sub-framework, of an argumentation framework $F = (A, R)$, such that $B^- \subseteq C$. It holds that:*

$$B \cup B^- \cup C = B \cup C \tag{5.2}$$

$$R_B \cup I_B \cup R_C = R_{B \cup C} \tag{5.3}$$

$$(B \cup C)^- = \emptyset \tag{5.4}$$

Proof. First, since $B^- \subseteq C$, it holds that: $B \cup B^- \cup C = B \cup C$, and $I_B = (C \times B) \cap R$. Second, since C is an unattacked set, $(B \times C) \cap R = ((B \cap C) \times C) \cap R \subseteq (C \times C) \cap R$, it follows that $R_B \cup I_B \cup R_C = ((B \times B) \cap R) \cup ((C \times B) \cap R) \cup ((C \times C) \cap R) = ((B \cup C) \times$

$(B \cup C)) \cap R = R_{B \cup C}$. Third, since $B^- \subseteq C$ and C is an unattacked set, it holds that $(B \cup C)^- = \emptyset$.

According to Proposition 5.3, the (syntactic) combination of $(B \cup B^-, R_B \cup I_B)$ and (C, R_C) is equal to $(B \cup C, R_{B \cup C})$, called the *combined sub-framework* of $(B \cup B^-, R_B \cup I_B)$ and (C, R_C). Semantically, we have the following definition.

Definition 5.4 (Combined extensions of a conditioned sub-framework and those of an unconditioned sub-framework).   Let $F = (A, R)$ be an argumentation framework, (C, R_C) and $(B \cup B^-, R_B \cup I_B)$ be the sub-frameworks of F, in which $B^- \subseteq C$. The set of *combined extensions* of $(B \cup C, R_{B \cup C})$, denoted as $CombExt_\sigma((B \cup C, R_{B \cup C}))$, is defined as follows:

$$CombExt_\sigma((B \cup C, R_{B \cup C})) = \{E_1 \cup E_2 \mid$$
$$E_1 \in \mathscr{E}_\sigma((C, R_C)) \wedge E_2 \in \mathscr{E}_\sigma((B \cup B^-, R_B \cup I_B)^{E_1})\}$$

Now, let us first verify the soundness of combining extensions of a conditioned sub-framework and those of an unconditioned sub-framework to form the extensions of a combined sub-framework (the completeness of this kind of combination will be presented at the end of this section).

Proposition 5.4.   *Let $F = (A, R)$ be an argumentation framework, (C, R_C) and $(B \cup B^-, R_B \cup I_B)$ be sub-frameworks of F, in which $B^- \subseteq C$. Under a semantics $\sigma \in \{adm, co, gr, pr\}$, for all $E \in CombExt_\sigma((B \cup C, R_{B \cup C}))$, it holds that $E \in \mathscr{E}_\sigma((B \cup C, R_{B \cup C}))$.*

Proof.

- Under admissible semantics, $E_1 \in \mathscr{E}_{ad}((C, R_C))$ and $E_2 \in \mathscr{E}_{ad}((B \cup B^-, R_B \cup I_B)^{E_1})$: in order to prove that $E_1 \cup E_2 \in \mathscr{E}_{ad}((B \cup C, R_{B \cup C}))$, we should prove that: (1) $E_1 \cup E_2$ is conflict-free, and (2) every argument in $E_1 \cup E_2$ is acceptable with respect to $E_1 \cup E_2$.

  (1) $\forall \alpha \in E_1 \subseteq C, \forall \beta \in E_2 \subseteq B$, there are the following four possible cases:

   (a) $\alpha \in C \setminus B$ and $\beta \in B \setminus C$: On the one hand, it holds that α does not attack β. Otherwise, assume that $(\alpha, \beta) \in R$, then β is attacked by a conditioning argument that is accepted with respect to E_1. According to the second condition of acceptability of arguments in a partially assigned sub-framework, β is not acceptable with respect to E_2 and $B^-[E_1]$, i.e., $\beta \notin E_2$, contradicting $\beta \in E_2$. On the other hand, β does not attack α. Otherwise, α is attacked by an argument outside C, contradicting the fact that C is an unattacked set.

   (b) $\alpha \in C \cap B$ and $\beta \in B \setminus C$: On the one hand, it holds that α does not attack β. Otherwise, since $\beta \in E_2$, there exists $\gamma \in E_2$ or $\gamma \in E_1$, such that γ attacks α. Since C is an unattacked set, $\gamma \notin B \setminus C$. It follow that $\gamma \in C$. As a result, α is attacked by an accepted argument in C, contradicting $\alpha \in E_1$. On the other hand, since C is an unattacked set and $\beta \in B \setminus C$, β does not attack α.

(c) $\alpha \in C \cap B$ and $\beta \in B \cap C$: On the one hand, it holds that α does not attack β (the reason is the same as the one in the previous item). On the other hand, β does not attack α. Otherwise, since both α and β are in C, and $\alpha \in E_1$, there exists $\gamma \in E_1$ such that γ attacks β, contradicting $\beta \in E_2$.

(d) $\alpha \in C \setminus B$ and $\beta \in B \cap C$: On the one hand, α does not attack β. Otherwise, β is attacked by a conditioning argument that is accepted with respect to E_1, and therefore $\beta \notin E_2$. Contradiction. On the other hand, β does not attack α (the reason is the same as the one in the previous item).

Since in all possible cases, neither α attacks β nor β attacks α, $E_1 \cup E_2$ is conflict-free.

(2) We should prove that $\forall \alpha \in E_1 \cup E_2, \forall \beta \in B \cup C$, if $(\beta, \alpha) \in R_{B \cup C}$, then $\exists \gamma \in E_1 \cup E_2$, such that $(\gamma, \beta) \in R_{B \cup C}$.

(a) $\forall \alpha \in E_1, \alpha \in C$. So, α is only attacked by the arguments in C (there exists no argument in $B \setminus C$ that attacks α). Since α is acceptable with respect to E_1, it holds that: $\forall \beta \in C \subseteq B \cup C$, if $(\beta, \alpha) \in R_C \subseteq R_{B \cup C}$, then $\exists \gamma \in E_1 \subseteq E_1 \cup E_2$, such that $(\gamma, \beta) \in R_C \subseteq R_{B \cup C}$.

(b) $\forall \alpha \in E_2 \setminus E_1$, there are two possible cases: α is attacked by the arguments in B^- with respect to I_B, or by the arguments in B with respect to R_B. So, if $(\beta, \alpha) \in R_{B \cup C}$, then it holds that: $\beta \in B^-$ and $(\beta, \alpha) \in I_B$, or $\beta \in B$ and $(\beta, \alpha) \in R_B$. Since α is acceptable with respect to E_2 and $B^-[E_1]$, it holds that:

(b1) if $\beta \in B^-$ and $(\beta, \alpha) \in I_B$, then β is rejected with respect to E_1 (corresponding to the second condition of acceptability of arguments in a partially assigned sub-framework), i.e., $\exists \gamma \in E_1 \subseteq E_1 \cup E_2$, such that $(\gamma, \beta) \in R_C \subseteq R_{B \cup C}$; and

(b2) if $\beta \in B$ and $(\beta, \alpha) \in R_B$, then $\exists \gamma \in E_2 \subseteq E_1 \cup E_2$, such that $(\gamma, \beta) \in R_B \subseteq R_{B \cup C}$, or $\exists \xi \in B^-$, such that ξ is accepted with respect to E_1 and $(\xi, \beta) \in I_B \subseteq R_{B \cup C}$, i.e., $\exists \xi \in E_1 \cap B^- \subseteq E_1 \cup E_2$, such that $(\xi, \beta) \in I_B \subseteq R_{B \cup C}$ (corresponding to the first condition of acceptability of arguments in a partially assigned sub-framework).

Putting (a) and (b) together, we have: $\forall \alpha \in E_1 \cup E_2, \forall \beta \in B \cup C$, if $(\beta, \alpha) \in R_{B \cup C}$, then $\exists \gamma \in E_1 \cup E_2$, such that $(\gamma, \beta) \in R_{B \cup C}$. Therefore, every argument in $E_1 \cup E_2$ is acceptable with respect to $E_1 \cup E_2$.

- Under complete semantics, $E_1 \in \mathscr{E}_{co}((C, R_C))$ and $E_2 \in \mathscr{E}_{co}((B \cup B^-, R_B \cup I_B)^{E_1})$: in order to prove that $E_1 \cup E_2 \in \mathscr{E}_{co}((B \cup C, R_{B \cup C}))$, we should prove that every argument in $B \cup C$ that is acceptable with respect to $E_1 \cup E_2$ is in $E_1 \cup E_2$. Assume the contrary, i.e. $\exists \alpha \in B \cup C$, such that α is acceptable with respect to $E_1 \cup E_2$, but $\alpha \notin E_1 \cup E_2$.

(1) If $\alpha \in C$, then α is only attacked by the arguments in C. Since α is acceptable with respect to $E_1 \cup E_2$, it holds that $\forall \beta \in C \subseteq B \cup C$ (there exists no argument in $B \setminus C$ that attacks α), if $(\beta, \alpha) \in R_{B \cup C} \cap (C \times C) = R_C$, then $\exists \gamma \in (E_1 \cup E_2) \cap C = E_1$, such that $(\gamma, \beta) \in R_{B \cup C} \cap (C \times C) = R_C$, i.e., α is acceptable with respect to E_1.

Since $\alpha \notin E_1 \cup E_2$, it holds that $\alpha \notin E_1$. So, E_1 is not a complete extension, contradicting $E_1 \in \mathcal{E}_{co}((C, R_C))$.

(2) If $\alpha \in B \setminus C$, then α is attacked by the arguments in B with respect to R_B or by the arguments in B^- with respect to I_B. Since α is acceptable with respect to $E_1 \cup E_2$, it holds that:

(a) $\forall \beta \in B$, if $(\beta, \alpha) \in R_B \subseteq R_{B \cup C}$, then: since β is in turn attacked by the arguments in B with respect to R_B or B^- with respect to I_B, it holds that $\exists \gamma \in (E_1 \cup E_2) \cap B = E_2$, such that $(\gamma, \beta) \in R_{B \cup C} \cap (B \times B) = R_B$, or $\exists \gamma' \in (E_1 \cup E_2) \cap B^- = E_1 \cap B^-$, such that $(\gamma', \beta) \in R_{B \cup C} \cap (B^- \times B) = I_B$ (satisfying the first condition of acceptability of arguments in a partially assigned sub-framework); and

(b) $\forall \beta \in B^-$, if $(\beta, \alpha) \in I_B \subseteq R_{B \cup C}$, then: since β is attacked by the arguments in C, it holds that $\exists \gamma \in (E_1 \cup E_2) \cap C = E_1$, such that $(\gamma, \beta) \in R_C \cap (C \times C) = R_C$, i.e., β is rejected with respect to E_1 (satisfying the second condition of acceptability of arguments in a partially assigned sub-framework).

As a result, α is acceptable with respect to E_2 and $B^-[E_1]$. Since $\alpha \notin E_1 \cup E_2$, it holds that $\alpha \notin E_2$. So, E_2 is not a complete extension, contradicting $E_2 \in \mathcal{E}_{co}((B \cup B^-, R_B \cup I_B)^{E_1})$.

According to (1) and (2), we have: $\forall \alpha \in B \cup C = C \cup B$, if α is acceptable with respect to $E_1 \cup E_2$, then $\alpha \in E_1 \cup E_2$. Therefore, every argument in $B \cup C$ that is acceptable with respect to $E_1 \cup E_2$ is in $E_1 \cup E_2$.

- Under preferred semantics, $E_1 \in \mathcal{E}_{pr}((C, R_C))$ and $E_2 \in \mathcal{E}_{pr}((B \cup B^-, R_B \cup I_B)^{E_1})$: since a preferred extension is also a complete extension, based on the proof in the case of complete semantics, we only need to prove that $E_1 \cup E_2$ is a maximal (with respect to set-inclusion) complete extension of $(B \cup C, R_{B \cup C})$. Assume that $E_1 \cup E_2$ is not a maximal complete extension. Then, $\exists S \subseteq C \cup B$, such that S is a preferred extension of $(B \cup C, R_{B \cup C})$ and $S \supseteq E_1 \cup E_2$. Let $\Phi = S \setminus (E_1 \cup E_2)$. If $\Phi = \emptyset$, then $S = E_1 \cup E_2$. In this case, $E_1 \cup E_2$ is a preferred extension. So, we need only to discuss the case when $\Phi \neq \emptyset$. Let $\Phi' = \Phi \cap C$ and $\Phi'' = \Phi \cap B$. It follows that $\Phi' \cup \Phi'' = \Phi$.

(1) When $\Phi' \neq \emptyset$, since $E_1 = (E_1 \cup E_2) \cap C$ and $\Phi' = \Phi \cap C$, it holds that $E_1 \cup \Phi' = (E_1 \cup E_2 \cup \Phi) \cap C = S \cap C$. According to Formula 5.1, it holds that $E_1 \cup \Phi' = S \cap C \in \mathcal{E}_{co}((C, R_C))$. Therefore, E_1 is not a preferred extension, contradicting $E_1 \in \mathcal{E}_{pr}((C, R_C))$.

(2) When $\Phi \neq \emptyset$, if $\Phi' = \emptyset$, then $\Phi'' \neq \emptyset$. In this case, since $E_2 = (E_1 \cup E_2) \cap B$ and $\Phi'' = \Phi \cap B$, it holds that $E_2 \cup \Phi'' = (E_1 \cup E_2 \cup \Phi) \cap B = S \cap B$. Since a preferred extension is also a complete extension, according to Proposition 5.1 and the proof of this proposition under complete semantics (in Section 5.2), it holds that $E_2 \cup \Phi'' = S \cap B \in \mathcal{E}_{co}((B \cup B^-, R_B \cup I_B)^{E_1})$. So, E_2 is not a preferred extension, contradicting $E_2 \in \mathcal{E}_{pr}((B \cup B^-, R_B \cup I_B)^{E_1})$.

As a result, we may conclude that $E_1 \cup E_2$ is a maximal complete extension (i.e., preferred extension).

- Under grounded semantics, $E_1 \in \mathscr{E}_{gr}((C, R_C))$ and $E_2 \in \mathscr{E}_{gr}((B \cup B^-, R_B \cup I_B)^{E_1})$: since a grounded extension is also a complete extension, based on the proof in the case of complete semantics, we only need to prove that $E_1 \cup E_2$ is the minimal (with respect to set-inclusion) complete extension of $(B \cup C, R_{B \cup C})$. The proof is similar to the one under preferred semantics (omitted).

Example 5.8.   Continue Example 5.6. Let $C = \{a_1, a_2, a_7\}$ and $B = \{a_2, a_3, a_4\}$. It follow that $B^- = a_1 \subseteq C$. Under complete semantics,

- $\mathscr{E}_{co}((C, R_C)) = \{E_1, E_2, E_3\}$, in which $E_1 = \{a_1, a_7\}$, $E_2 = \{a_2\}$ and $E_3 = \emptyset$;
- $\mathscr{E}_{co}((B \cup B^-, R_B \cup I_B)^{E_1}) = \{E_4\}$, in which $E_4 = \{a_1, a_3\}$;
- $\mathscr{E}_{co}((B \cup B^-, R_B \cup I_B)^{E_2}) = \{E_5\}$, in which $E_5 = \{a_2, a_4\}$;
- $\mathscr{E}_{co}((B \cup B^-, R_B \cup I_B)^{E_3}) = \{E_6\}$, in which $E_6 = \emptyset$.

According to Definition 5.4, we have $Comb Ext_\sigma((B \cup C, R_{B \cup C})) = \{E_7, E_8, E_9\}$, in which

- $E_7 = E_1 \cup E_4 = \{a_1, a_3, a_7\}$,
- $E_8 = E_2 \cup E_5 = \{a_2, a_4\}$, and
- $E_8 = E_3 \cup E_6 = \emptyset$.

Based on Proposition 5.4, firstly, let us now prove the two remaining parts of Proposition 5.1 mentioned in Section 5.2.

Proof.

- In the first part, we prove the soundness of restricting an extension to a conditioned sub-framework under preferred and grounded semantics, i.e., for all $\sigma \in \{pr, gr\}$, if $E \in \mathscr{E}_\sigma(F)$, then $E \cap (B \cup B^-)$ is an extension of $(B \cup B^-, R_B \cup I_B)^{E_1}$. According to the proof in Section 5.2, $E \cap (B \cup B^-) \in \mathscr{E}_{co}((B \cup B^-, R_B \cup I_B)^{E_1})$, in which $E_1 = E \cap C$. Based on this result, we have the following proof.

  (1) Under preferred semantics, we only need to verify that $E \cap (B \cup B^-)$ is maximal. Assume the contrary. Then, $\exists E_2 \in \mathscr{E}_{pr}((B \cup B^-, R_B \cup I_B)^{E_1})$, such that $E_2 \supset E \cap (B \cup B^-)$. It follows that $E_1 \cup E_2 \supset E_1 \cup (E \cap (B \cup B^-)) = (E \cap C) \cup (E \cap (B \cup B^-)) = E \cap (B \cup C)$. According to Formula 5.1, $E \cap (B \cup C)$ is a preferred extension of $(B \cup C, R_{B \cup C})$, and $E \cap C$ is a preferred extension of (C, R_C). Since $E_1 \in \mathscr{E}_{pr}((C, R_C))$ and $E_2 \in \mathscr{E}_{pr}((B \cup B^-, R_B \cup I_B)^{E_1})$, according to Proposition 5.4, $E_1 \cup E_2$ is a preferred extension of $(B \cup C, R_{B \cup C})$. As a result, $E \cap (B \cup C)$ is not a preferred extension of $(B \cup C, R_{B \cup C})$. Contradiction.
  (2) Under grounded semantics, we only need to verify that $E \cap (B \cup B^-)$ is minimal. The proof is similar to the previous item. Omitted.

Figure 5.6 Argumentation framework $F_{5.5}$.

- In the second part, we prove the completeness of restricting an extension to a conditioned sub-framework under all four semantics, i.e., for all $\sigma \in \{adm, co, pr, gr\}$, if E_2 is an extension of $(B \cup B^-, R_B \cup I_B)^{E_1}$, then $\exists E \in \mathcal{E}_\sigma(F)$ such that $E \cap (B \cup B^-) = E_2$. According to Proposition 5.4, $E_1 \cup E_2$ is an extension of $(B \cup C, R_{B \cup C})$. Let $D = A \setminus (B \cup C)$. It holds that $D^- \subseteq B \cup C$. For all $E_3 \in \mathcal{E}_\sigma((D \cup D^-, R_D \cup I_D)^{E_1 \cup E_2})$, $E_1 \cup E_2 \cup E_3 \in \mathcal{E}_\sigma(F)$. Let $E = E_1 \cup E_2 \cup E_3$. It holds that $E_2 = E \cap (B \cup B^-)$.

According to the soundness and completeness of restricting an extension to a conditioned sub-framework, with respect to a conditioned sub-framework, the mapping from global semantics to local semantics can also be formulated by the following definition:

Definition 5.5 (Mapping global semantics to local semantics with respect to a conditioned sub-framework). Let $F = (A, R)$ be an argumentation framework, and (C, R_C) and $(B \cup B^-, R_B \cup I_B)$ be sub-frameworks of F. Under a semantics $\sigma \in \{adm, co, pr, gr\}$, for all $E \in \mathcal{E}_\sigma(F)$, it holds that

$$\mathcal{E}_\sigma((B \cup B^-, R_B \cup I_B)^{E \cap C}) = \{E' \cap (B \cup B^-) \mid E' \in \mathcal{E}_\sigma(F) \wedge E' \cap C = E \cap C\} \quad (5.5)$$

Example 5.9. Let us consider an argumentation framework $F_{5.5} = (A, R)$ as shown in Figure 5.6. Let $B = \{a_3, a_4, a_5, a_6\}$ and $C = \{a_1, a_2\}$. It follows that $B^- = \{a_2\}$. Under preferred semantics, $F_{5.5}$ has the following four extensions:

- $E_1 = \{a_1, a_3, a_5\}$,
- $E_2 = \{a_1, a_3, a_6\}$,
- $E_3 = \{a_1, a_4, a_6\}$, and
- $E_4 = \{a_2, a_4, a_6\}$.

It holds that $E_1 \cap C = E_2 \cap C = E_3 \cap C = \{a_1\}$, and $E_4 \cap C = \{a_2\}$. According to Formula 5.5, we have

- $\mathcal{E}_{pr}((B \cup B^-, R_B \cup I_B)^{\{a_1\}}) = \{E_5, E_6, E_7\}$, in which
  - $E_5 = E_1 \cap (B \cup B^-) = \{a_3, a_5\}$,
  - $E_6 = E_2 \cap (B \cup B^-) = \{a_3, a_6\}$, and
  - $E_7 = E_3 \cap (B \cup B^-) = \{a_4, a_6\}$.
- $\mathcal{E}_{pr}((B \cup B^-, R_B \cup I_B)^{\{a_2\}}) = \{E_8\}$, in which $E_8 = E_4 \cap (B \cup B^-) = \{a_2, a_4, a_6\}$.

Finally, according to the soundness of restricting an extension to a sub-framework (conditioned or unconditioned), we may verify the completeness of combining extensions of a conditioned sub-framework and those of an unconditioned sub-framework to form the extensions of a combined sub-framework.

Proposition 5.5. *Let $F = (A, R)$ be an argumentation framework, (C, R_C) and $(B \cup B^-, R_B \cup I_B)$ be sub-frameworks of F, in which $B^- \subseteq C$. Under a semantics $\sigma \in \{adm, co, gr, pr\}$, for all $E \in \mathscr{E}_\sigma((B \cup C, R_{B \cup C}))$, it holds that $E \in CombExt_\sigma((B \cup C, R_{B \cup C}))$.*

Proof. Under a given semantics $\sigma \in \{adm, co, gr, pr\}$, for all $E \in \mathscr{E}_\sigma((B \cup C, R_{B \cup C}))$, let $E_1 = C \cap E$ and $E_2 = (B \cup B^-) \cap E$. It holds that $E_1 \cup E_2 = E \cap (B \cup C \cup B^-) = E$. According to Formulas 5.1 and 5.5, it holds that $E_1 \in \mathscr{E}_\sigma((C, R_C))$ and $E_2 \in \mathscr{E}_\sigma((B \cup B^-, R_B \cup I_B)^{E_1})$. According to Definition 5.4, it holds that $E_1 \cup E_2 = E \in CombExt_\sigma((B \cup C, R_{B \cup C}))$.

5.3.3 Combining Labellings of Two Conditioned Sub-Frameworks

In the previous two subsections, by using the extension-based approach, we have presented the semantics combination of two unconditioned sub-frameworks, and of a conditioned sub-framework and an unconditioned sub-framework. In this subsection, we will formulate the semantics combination of two conditioned sub-frameworks, in terms of labelling-based approach.

Let $F = (A, R)$ be an argumentation framework, and $(B_1 \cup B_1^-, R_{B_1} \cup I_{B_1})$ and $(B_2 \cup B_2^-, R_{B_2} \cup I_{B_2})$ be two conditioned sub-frameworks. The (syntactic) combination of the two sub-frameworks is $(B_1 \cup B_2 \cup B_1^- \cup B_2^-, R_{B_1} \cup R_{B_2} \cup I_{B_1} \cup I_{B_2})$.

There are some possible relations between B_1 and B_2, B_1^- and B_2, and B_1 and B_2^-. In this book, we only consider the case where $B_1 \cap B_2 = \emptyset$, $B_1^- \cap B_2 = \emptyset$ and $B_1 \cap B_2^- = \emptyset$.

Let $B = B_1 \cup B_2$. According to the following proposition, $(B_1 \cup B_2 \cup B_1^- \cup B_2^-, R_{B_1} \cup R_{B_2} \cup I_{B_1} \cup I_{B_2})$ is equal to $(B \cup B^-, R_B \cup I_B)$.

Proposition 5.6. *Let $(B_1 \cup B_1^-, R_{B_1} \cup I_{B_1})$ and $(B_2 \cup B_2^-, R_{B_2} \cup I_{B_2})$ be conditioned sub-frameworks of $F = (A, R)$, such that $B_1 \cap B_2 = \emptyset$, $B_1^- \cap B_2 = \emptyset$ and $B_1 \cap B_2^- = \emptyset$. Let $B = B_1 \cup B_2$. It holds that:*

(1) $B^- = B_1^- \cup B_2^-$,
(2) $R_B = R_{B_1} \cup R_{B_2}$, *and*
(3) $I_B = I_{B_1} \cup I_{B_2}$.

Proof. (1) On the one hand, for all $\alpha \in B^-$, there exists $\beta \in B$ such that $(\alpha, \beta) \in R$. Since $B = B_1 \cup B_2$, we have the following two possible cases. When $\beta \in B_1$, since $\alpha \notin B_1$, $\alpha \in A \setminus B_1$. It follows that $\alpha \in B_1^-$. Similarly, when $\beta \in B_2$, since $\alpha \notin B_2$, $\alpha \in A \setminus B_2$. Hence, $\alpha \in B_1^- \cup B_2^-$. On the other hand, for all $\alpha \in B_1^- \cup B_2^-$, $\alpha \in B_1^-$ or $\alpha \in B_2^-$. When

$\alpha \in B_1^-$, there exists $\beta \in B_1$ such that $(\alpha, \beta) \in R$. Since $B_1^- \cap B_2 = \emptyset$, $\alpha \notin B_2$. It follows that $\alpha \notin B_1 \cup B_2$ and $\alpha \in A \setminus (B_1 \cup B_2)$. So, $\alpha \in (B_1 \cup B_2)^-$. Similarly, when $\alpha \in B_2^-$, it holds that $\alpha \in (B_1 \cup B_2)^-$.

(2) Since $B_1^- \cap B_2 = \emptyset$ and $B_1 \cap B_2^- = \emptyset$, it holds that there are no interactions between B_1 and B_2, i.e., $R \cap (B_1 \times B_2) = \emptyset$ and $R \cap (B_2 \times B_1) = \emptyset$. So,
$R_B = R \cap ((B_1 \cup B_2) \times (B_1 \cup B_2)) = (R \cap (B_1 \times B_1)) \cup (R \cap (B_2 \times B_2)) = R_{B_1} \cup R_{B_2}$.

(3) Since $B_1^- \cap B_2 = \emptyset$ and $B_1 \cap B_2^- = \emptyset$, $I_B = R \cap (B^- \times B) =$
$R \cap ((B_1^- \cup B_2^-) \times (B_1 \cup B_2)) = (R \cap (B_1^- \times B_1)) \cup (R \cap (B_2^- \times B_2)) = I_{B_1} \cup I_{B_2}$.

Now, let us define the combination of the labellings of two conditioned sub-frameworks. Let $\mathcal{L}_1 = (in(\mathcal{L}_1), out(\mathcal{L}_1), undec(\mathcal{L}_1))$ and $\mathcal{L}_2 = (in(\mathcal{L}_2), out(\mathcal{L}_2), undec(\mathcal{L}_2))$ be two labellings. The combination of $\mathcal{L}_1$ and $\mathcal{L}_2$ is denoted as:

$$\mathcal{L}_1 + \mathcal{L}_2 = (in(\mathcal{L}_1) \cup in(\mathcal{L}_2), out(\mathcal{L}_1) \cup out(\mathcal{L}_1), undec(\mathcal{L}_1) \cup undec(\mathcal{L}_1)) \qquad (5.6)$$

Based on this notion, we have the following definition.

Definition 5.6 (Combined labellings of two conditioned sub-frameworks). Let $(B_1 \cup B_1^-, R_{B_1} \cup I_{B_1})$ and $(B_2 \cup B_2^-, R_{B_2} \cup I_{B_2})$ be conditioned sub-frameworks of $F = (A, R)$, such that $B_1 \cap B_2 = \emptyset$, $B_1^- \cap B_2 = \emptyset$ and $B_1 \cap B_2^- = \emptyset$. Let $B = B_1 \cup B_2$. So, $(B \cup B^-, R_B \cup I_B)$ is a combined framework of $(B_1 \cup B_1^-, R_{B_1} \cup I_{B_1})$ and $(B_2 \cup B_2^-, R_{B_2} \cup I_{B_2})$. Let (C_1, R_{C_1}) and (C_2, R_{C_2}) be two unconditioned sub-frameworks of F, such that $B_1^- \subseteq C_1$ and $B_2^- \subseteq C_2$. Let σ be a semantics, under which every argumentation framework has at least one labelling. For all $\mathcal{L}_1 \in \mathcal{L}_\sigma((C_1, R_{C_1}))$, $\mathcal{L}_2 \in \mathcal{L}_\sigma((C_2, R_{C_2}))$, let $\mathcal{L} = \mathcal{L}_1 + \mathcal{L}_2$.

$$CombLab_\sigma((B \cup B^-, R_B \cup I_B)^{\mathcal{L}}) = \{\mathcal{L}'_1 + \mathcal{L}'_2 \mid$$
$$\mathcal{L}'_1 \in \mathcal{L}_\sigma((B_1 \cup B_1^-, R_{B_1} \cup I_{B_1})^{\mathcal{L}_1})$$
$$\wedge \mathcal{L}'_2 \in \mathcal{L}_\sigma((B_2 \cup B_2^-, R_{B_2} \cup I_{B_2})^{\mathcal{L}_2})\}$$

The soundness and completeness of combining labellings of two conditioned sub-frameworks are formulated by the following proposition.

Proposition 5.7. *Based on Definition 5.6, it holds that*
$\mathcal{L}_\sigma((B \cup B^-, R_B \cup I_B)^{\mathcal{L}}) = CombLab_\sigma((B \cup B^-, R_B \cup I_B)^{\mathcal{L}})$.

Proof. $\Leftarrow$ (Soundness): For all $\mathcal{L}'_1 \in \mathcal{L}_\sigma((B_1 \cup B_1^-, R_{B_1} \cup I_{B_1})^{\mathcal{L}_1})$ and $\mathcal{L}'_2 \in \mathcal{L}_\sigma((B_2 \cup B_2^-, R_{B_2} \cup I_{B_2})^{\mathcal{L}_2})$, it holds that $\mathcal{L}'_1 + \mathcal{L}'_2 \in \mathcal{L}_\sigma((B \cup B^-, R_B \cup I_B)^{\mathcal{L}})$.

(1) Under admissible semantics, $\mathcal{L}_1$ and $\mathcal{L}_2$ are admissible labellings. According to Propositions 5.2 and 2.1, $\mathcal{L} = \mathcal{L}_1 + \mathcal{L}_2$ is an admissible labelling. For all

$\mathcal{L}'_1 \in \mathcal{L}_{adm}((B_1 \cup B_1^-, R_{B_1} \cup I_{B_1})^{\mathcal{L}_1})$ and $\mathcal{L}'_2 \in \mathcal{L}_{adm}((B_2 \cup B_2^-, R_{B_2} \cup I_{B_2})^{\mathcal{L}_2})$, we need to verify that $\mathcal{L}'_1 + \mathcal{L}'_2$ is an admissible labelling of $(B \cup B^-, R_B \cup I_B)^{\mathcal{L}}$. According to Definition 4.6, if for $l \in \{IN, OUT\}$, each argument α in B that is labelled l in $\mathcal{L}'_1 + \mathcal{L}'_2$ is legally l in $\mathcal{L}'_1 + \mathcal{L}'_2$ with regard to $\mathcal{L}$, then $\mathcal{L}'_1 + \mathcal{L}'_2$ is an admissible labelling with respect to $\mathcal{L}$.

Since $\alpha \in B = B_1 \cup B_2$, and $B_1 \cap B_2 = \emptyset$, it holds that $\alpha \in B_i$ (i is either 1 or 2). Since α is labelled l in $\mathcal{L}'_1 + \mathcal{L}'_2$ and $\alpha \in B_i$, α is labelled l in $\mathcal{L}'_i$. Since $\mathcal{L}'_i$ is an admissible labelling, it holds that α is legally l in $\mathcal{L}'_i$ with regard to $\mathcal{L}_i$. Since α is only possibly attacked by the arguments in B_i and B_i^-, it holds that α is legally l in $\mathcal{L}'_1 + \mathcal{L}'_2$ with regard to $\mathcal{L}_1 + \mathcal{L}_2 = \mathcal{L}$.

(2) Under complete semantics, $\mathcal{L}_1$ and $\mathcal{L}_2$ are complete labellings. According to Propositions 5.2 and 2.2, $\mathcal{L} = \mathcal{L}_1 + \mathcal{L}_2$ is a complete labelling. For all $\mathcal{L}'_1 \in \mathcal{L}_{co}((B_1 \cup B_1^-, R_{B_1} \cup I_{B_1})^{\mathcal{L}_1})$ and $\mathcal{L}'_2 \in \mathcal{L}_{co}((B_2 \cup B_2^-, R_{B_2} \cup I_{B_2})^{\mathcal{L}_2})$, we need to verify that $\mathcal{L}'_1 + \mathcal{L}'_2$ is a complete labelling of $(B \cup B^-, R_B \cup I_B)^{\mathcal{L}}$. Since a complete labelling is an admissible labelling, according to Definition 4.6, we only need to verify that each argument α in B that is labelled UNDEC in $\mathcal{L}'_1 + \mathcal{L}'_2$ is legally UNDEC in $\mathcal{L}'_1 + \mathcal{L}'_2$ with regard to $\mathcal{L}$. Since $\alpha \in B_i$ (i is either 1 or 2) and α is labelled UNDEC in $\mathcal{L}'_1 + \mathcal{L}'_2$, α is legally UNDEC in $\mathcal{L}'_i$ with regard to $\mathcal{L}_i$ ($i = 1$ or 2). Since α is only possibly attacked by the arguments in B_i and B_i^-, it holds that α is legally UNDEC in $\mathcal{L}'_1 + \mathcal{L}'_2$ with regard to $\mathcal{L}_1 + \mathcal{L}_2 = \mathcal{L}$.

(3) Under preferred semantics, $\mathcal{L}_1$ and $\mathcal{L}_2$ are preferred labellings. According to Propositions 5.2 and 2.2, $\mathcal{L} = \mathcal{L}_1 + \mathcal{L}_2$ is a preferred labelling. For all $\mathcal{L}'_1 \in \mathcal{L}_{pr}((B_1 \cup B_1^-, R_{B_1} \cup I_{B_1})^{\mathcal{L}_1})$ and $\mathcal{L}'_2 \in \mathcal{L}_{pr}((B_2 \cup B_2^-, R_{B_2} \cup I_{B_2})^{\mathcal{L}_2})$, we need to verify that $\mathcal{L}'_1 + \mathcal{L}'_2$ is a preferred labelling of $(B \cup B^-, R_B \cup I_B)^{\mathcal{L}}$. Since a preferred labelling is also a complete labelling, based on the proof in the case of complete semantics, we only need to prove that $in(\mathcal{L}'_1 + \mathcal{L}'_2)$ is maximal.

Assume the contrary. Then, there exists another complete labelling $\mathcal{L}'$ of $(B \cup B^-, R_B \cup I_B)^{\mathcal{L}}$, such that $in(\mathcal{L}') \supset in(\mathcal{L}'_1 + \mathcal{L}'_2)$. Let $\Phi = in(\mathcal{L}') \setminus in(\mathcal{L}'_1 + \mathcal{L}'_2)$. It holds that $\Phi \neq \emptyset$.

Let $\Phi_1 = \Phi \cap (B_1 \cup B_1^-)$ and $\Phi_2 = \Phi \cap (B_2 \cup B_2^-)$. Since $\Phi_1 \cup \Phi_2 = \Phi \cap (B \cup B^-) = \Phi \neq \emptyset$, it holds that $\Phi_1 \neq \emptyset$ or $\Phi_2 \neq \emptyset$. Without loss of generality, let $\Phi_1 \neq \emptyset$. Since $in(\mathcal{L}'_1) \cap (B_1 \cup B_1^-) \subseteq B_1 \cup B_1^-$, $in(\mathcal{L}'_2) \cap B_1 = \emptyset$, and $in(\mathcal{L}'_2) \cap B_1^- \subseteq in(\mathcal{L}'_1 + \mathcal{L}'_2) \cap B_1^- = in(\mathcal{L}'_1) \cap B_1^-$, it holds that $in(\mathcal{L}'_2) \cap (B_1 \cup B_1^-) = (in(\mathcal{L}'_1) \cup in(\mathcal{L}'_2)) \cap (B_1 \cup B_1^-) = in(\mathcal{L}'_1 + \mathcal{L}'_2) \cap (B_1 \cup B_1^-)$. So, $in(\mathcal{L}'_1) \cup \Phi_1 = (in(\mathcal{L}'_1 + \mathcal{L}'_2) \cap (B_1 \cup B_1^-)) \cup (\Phi \cap (B_1 \cup B_1^-)) = (in(\mathcal{L}'_1 + \mathcal{L}'_2) \cup \Phi) \cap (B_1 \cup B_1^-) = in(\mathcal{L}') \cap (B_1 \cup B_1^-)$.

According to Formula 5.5 and Proposition 2.2, we may infer that $(in(\mathcal{L}') \cap (B_1 \cup B_1^-), out(\mathcal{L}') \cap (B_1 \cup B_1^-), undec(\mathcal{L}') \cap (B_1 \cup B_1^-))$ is a complete labelling of $(B_1 \cup B_1^-, R_{B_1} \cup I_{B_1})^{\mathcal{L}_1}$. Since $in(\mathcal{L}') \cap (B_1 \cup B_1^-) = in(\mathcal{L}'_1) \cup \Phi_1 \supset in(\mathcal{L}'_1)$, $\mathcal{L}'_1$ is not a preferred labelling, contradicting $\mathcal{L}'_1 \in \mathcal{L}_{pr}((B_1 \cup B_1^-, R_{B_1} \cup I_{B_1})^{\mathcal{L}_1})$.

(4)   Under grounded semantics, $\mathcal{L}_1$, $\mathcal{L}_2$ and $\mathcal{L}$ are grounded labellings. For all
$\mathcal{L}_1' \in \mathscr{L}_{gr}((B_1 \cup B_1^-, R_{B_1} \cup I_{B_1})^{\mathcal{L}_1})$ and $\mathcal{L}_2' \in \mathscr{L}_{gr}((B_2 \cup B_2^-, R_{B_2} \cup I_{B_2})^{\mathcal{L}_2})$, we need to
verify that $\mathcal{L}_1' + \mathcal{L}_2'$ is the grounded labelling of $(B \cup B^-, R_B \cup I_B)^{\mathcal{L}}$. Since a grounded
labelling is also a complete labelling, based on the proof in the case of complete
semantics, we only need to prove that $in(\mathcal{L}_1' + \mathcal{L}_2')$ is minimal. The proof is similar to the
one in the case of preferred semantics (omitted).

$\Rightarrow$ (Completeness): For every labelling $\mathcal{L}' \in \mathscr{L}_\sigma((B \cup B^-, R_B \cup I_B)^{\mathcal{L}})$, it holds that $\mathcal{L}'$ is a
combined labelling of $(B \cup B^-, R_B \cup I_B)^{\mathcal{L}}$, i.e., $\mathcal{L}' \in CombLab_\sigma((B \cup B^-, R_B \cup I_B)^{\mathcal{L}})$.

Let $\mathcal{L}_j' = (in(\mathcal{L}') \cap (B_j \cup B_j^-), out(\mathcal{L}') \cap (B_j \cup B_j^-), undec(\mathcal{L}') \cap (B_j \cup B_j^-))$, for all
$j \in \{1, 2\}$. According to Formula 5.5 and Proposition 2.2, we may infer that
$\mathcal{L}_j' \in \mathscr{L}_\sigma((B_j \cup B_j^-, R_{B_j} \cup I_{B_j})^{\mathcal{L}_j})$. Since $\mathcal{L}_1' + \mathcal{L}_2' = \mathcal{L}'$, according to Definition 5.6, it holds
that $\mathscr{L}' \in CombLab_\sigma((B \cup B^-, R_B \cup I_B)^{\mathcal{L}})$.

5.4 Conclusions

This chapter has presented the relations between global semantics and local semantics of an
argumentation framework, which lays a foundation for efficient computation of argumentation
semantics. The notions related to the global semantics and local semantics also appeared in
[2,3] and [4]. In addition, this notion may be extended to the context of multiple argumentation
systems. In [5], the authors proposed a notion of merging (different) argumentation systems.
Given a set of distinct argumentation frameworks from different agents, they are expanded
respectively into partial systems over the set of all arguments considered by the group of
agents. Then, a *merging* operator is used to produce a set of argumentation systems that are as
close as possible to the partial systems (to realize a kind of consensus). And then, the
acceptability of a set of arguments at the group level is obtained by selecting the extensions of
a set of produced (merged) argumentation frameworks at the local level.

References

[1] P. Baroni, M. Giacomin, On principle-based evaluation of extension-based argumentation semantics, Artificial
     Intelligence 171 (10–15) (2007) 675–700.
[2] R. Baumann, Splitting an argumentation framework, in: Proceedings of the 11th International Conference on
     Logic Programming and Nonmonotonic Reasoning, 2011, pp. 40–53.
[3] R. Baumann, G. Brewka, R. Wong, Splitting argumentation frameworks: an empirical evaluation, in:
     Proceedings of the 1st International Workshop on Theory and Applications of Formal, Argumentation, 2011,
     pp. 17–31.
[4] V. Lifschitz, H. Turner, Splitting a Logic Program, Principles of Knowledge Representation (1994) 23–37.
[5] S. Coste-Marquis, C. Devred, S. Konieczny, M. Lagasquie-Schiex, P. Marquis, On the merging of Dung's
     argumentation systems, Artificial Intelligence 171 (10–15) (2007) 730–753.

CHAPTER 6

An Approach for Static Argumentation Frameworks

6.1 Introduction

As introduced in Chapter 1, due to the fact that many natural questions regarding argument acceptability are computationally intractable [1], how to efficiently compute the semantics of an argumentation framework is still an open problem.

Since the tractability of an argumentation framework is closely related to its structure [1], it is intuitively feasible to develop efficient computation methods by taking advantage of the special topologies of argumentation frameworks. Indeed, some tractable classes of argumentation frameworks have been clearly identified so far, including acyclic argumentation frameworks [2], symmetric argumentation frameworks [3], and bounded tree-width argumentation frameworks [1,4], etc. However, for an argumentation framework that may not belong to any tractable class, how to efficiently compute its semantics is still unexplored.

Efficient Computation of Argumentation Semantics. http://dx.doi.org/10.1016/B978-0-12-410406-8.00006-3

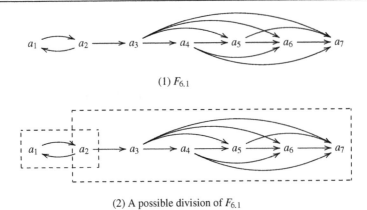

(1) $F_{6.1}$

(2) A possible division of $F_{6.1}$

Figure 6.1   An argumentation framework and a possible division of it.

In this chapter, inspired by the local tractability of an argumentation framework, we introduce an incremental approach to compute the semantics of a static argumentation framework. The basic idea is as follows:

The existing proof theories and algorithms [2,5–12] all treat argumentation frameworks as monolithic entities[1]. As a result, the tractability of various parts of a generic argumentation framework (called *local tractability*) has not been exploited. So, if it is feasible to decompose an argumentation framework into a set of sub-frameworks and compute the semantics of each sub-framework locally, then the local tractability could be exploited.

According to this intuitive observation, we naturally have the following hypothesis:

Given a generic argumentation framework, if we consider its topology as a whole, it may not belong to any tractable class. However, if we decompose it into a set of sub-frameworks, some of them may be tractable, while others might not, but have smaller sizes. This idea could be illustrated by the following example.

Example 6.1.   Let $F_{6.1} = (A, R)$ be an argumentation framework, in which $A = \{a_1, \ldots, a_n\}$, $R = \{(a_1, a_2), (a_2, a_1), (a_2, a_3)\} \cup \{(a_i, a_j) \mid 3 \le i < j \le n\}$, $n \ge 4$. For $n = 7$, $F_{6.1}$ is shown in Figure 6.1(1).

According to [4], the tree-width of $F_{6.1}$ (with increasing n) cannot be bounded by a constant. Meanwhile, $F_{6.1}$ is neither acyclic nor symmetric. So, $F_{6.1}$ may not belong to any tractable class. However, after we decompose $F_{6.1}$ into two sub-frameworks ($\{a_1, a_2\}, \{(a_1, a_2), (a_2, a_1)\}$) and ($\{a_2, \ldots, a_7\}, R \setminus \{(a_1, a_2), (a_2, a_1)\}$), it holds that the former is symmetric, and the latter is acyclic. So, if it is feasible to compute the semantics of the two

---

[1]   In some existing literature [13,15], argumentation frameworks are not treated as monolithic entities. However, their work is not focused on developing efficient methods to compute argumentation semantics.

sub-frameworks locally and combine them to form the semantics of $F_{6.1}$, then the overall time complexity of computation will be reduced.

Based on the above idea, we hope that the semantics of an argumentation framework could be computed by means of: (i) decomposing the argumentation framework into a set of sub-frameworks; (ii) computing the semantics of each sub-framework locally; and (iii) combining the semantics of a set of sub-frameworks to form the semantics of the original argumentation framework, such that:

(a) the local tractability of the argumentation framework could be exploited to some extent; and
(b) the size of each intractable sub-framework could be made as small as possible.

In order to exploit the local tractability of an argumentation framework, we need to identify tractable fragments. In this book, for simplicity, we only consider acyclic fragments, while the handling of other classes of fragments is left to future work.

6.2 Decomposing an Argumentation Framework: A Layered Approach

In order to enable the incremental computation of argumentation semantics, we need to decompose an argumentation framework in a general way such that there exists a partial order among different sub-frameworks with respect to the dependent relation. As explained in Example 4.3, this requirement might not be satisfied, if an argumentation framework is decomposed in an arbitrary manner. In this example, it turns out that all sub-frameworks are interdependent. The basic reason behind it is that some arguments in these sub-frameworks are in single strongly connected components (SCCs). Now, a natural question arises: is there a partial order among different SCCs? According to graph theory, the answer is positive. Meanwhile, since the set of SCCs of an argumentation framework can be obtained by a polynomial time algorithm [14], it is intuitively feasible to decompose an argumentation framework based on its SCCs.

6.2.1 Strongly Connected Components of an Argumentation Framework

The notion of *strongly connected components* (SCCs) of an argumentation framework is directly from graph theory, and has been reformulated in [15].

Definition 6.1 (Strongly connected components). Given an argumentation framework $F = (A, R)$, the set of strongly connected components of F (denoted as $SCCS_F$) are the equivalence classes of nodes under the relation of path-equivalence between nodes, which, denoted as $PE_F \subseteq (A \times A)$, is defined as follows:

(i) $\forall \alpha \in A, (\alpha, \alpha) \in PE_F$, and

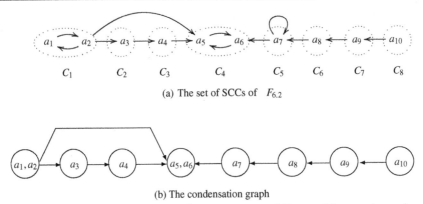

(a) The set of SCCs of $F_{6.2}$

(b) The condensation graph

Figure 6.2 The set of strongly connected components of $F_{6.2}$ and its condensation graph.

(ii)   given two distinct nodes $\alpha, \beta \in A$, $(\alpha, \beta) \in PE_F$, if and only if there is a path from α to β and a path from β to α.

According to graph theory, an important property of SCCs is that *every directed graph corresponds to a directed acyclic graph of its SCCs*. In other words, if we shrink each SCC down to a single meta-node, and draw an edge from one meta-node to another if there is an edge (in the same direction) between their respective components, then the resulting condensation graph must be a directed acyclic graph [16].

Example 6.2. Let $F_{6.2} = (A, R)$ be an argumentation framework, where:

$$A = \{a_1, \ldots, a_{10}\}$$
$$R = \{(a_1, a_2), (a_2, a_1), (a_2, a_3), (a_2, a_5), (a_3, a_4), (a_4, a_5), (a_5, a_6),$$
$$(a_6, a_5), (a_7, a_6), (a_7, a_7), (a_8, a_7), (a_9, a_8), (a_{10}, a_9)\}$$

According to Definition 6.1, $F_{6.2}$ is decomposed into eight SCCs: $C_1, \ldots, C_8$, i.e., $SCCS_{F_{6.2}} = \{C_1, \ldots, C_8\}$ (Figure 6.2(a)). The corresponding condensation graph (Figure 6.2(b)) is acyclic.

6.2.2 A Decomposition Approach Based on SCCs

Since there exists a partial order among different SCCs, a simple approach to decompose an argumentation framework is to use each SCC to induce a sub-framework. However, for each acyclic sub-framework (which is polynomial time tractable), it is induced by the union of a set of SCCs, each of which contains a single argument. Let us consider again the argumentation framework $F_{6.1}$ in Example 6.1, the sub-framework induced by the set $\{a_3, a_4, a_5, a_6, a_7\}$ is composed of five SCCs, each of which only contains a single argument. For a SCC containing a single argument which does not self-attack, we call it a trivial SCC. Otherwise, it is non-trivial.

Definition 6.2 (Trivial/non-trivial SCCs). Let $F = (A, R)$ be an argumentation framework, and SCCS_F be the set of SCCs of F. For all $C \in \mathrm{SCCS}_F$, if there is only one argument (say α) in C and the only argument does not self-attack, i.e., $(\alpha, \alpha) \notin R$, then we call C a *trivial* SCC, else a *non-trivial* SCC.

With the notion of trivial SCCs, acyclic sub-frameworks could be constructed by using the unions of trivial SCCs, as long as the partial order of dependence relation among different sub-frameworks is preserved. Meanwhile, since the sub-framework induced by each non-trivial SCC contains cycles, it might be intractable. So, we simply use a single non-trivial SCC to induce a sub-framework.

Besides the considerations for constructing sub-frameworks, when decomposing an argumentation framework, we need also to consider how to organize the sub-frameworks, such that their semantics can be computed incrementally. According to the partial order over the set of SCCs, it is natural to adopt a *layered approach*. The basic idea is that the sub-frameworks in the same layer can be handled simultaneously, while the sub-frameworks in a given layer are only dependent on the sub-frameworks in the lower layers. And, in the lowest layer, all sub-frameworks are independent.

In order to realize a layered decomposition approach, given the set of SCCs of an argumentation framework, we assign to each SCC a natural number, called the *level* of the SCC, indicating the order of computing the status of arguments in this SCC, such that:

(i) each non-trivial SCC is used to induce a sub-framework, and
(ii) the union of a set of trivial SCCs is used to induce an acyclic sub-framework, under the condition that the partial order among different sub-frameworks is preserved.

Formally, the notion of the *level* of a SCC is defined as follows.

Definition 6.3 (Level of a SCC). Let SCCS_F be the set of SCCs of an argumentation framework $F = (A, R)$. For all $C \in \mathrm{SCCS}_F$, the *level* of C is a function $\rho : \mathrm{SCCS}_F \to \mathbb{N}$. Recursively, $\rho(C)$ is defined as follows:

1. If C has no parent, then $\rho(C) = 0$;
2. Else:
   (a) when C is non-trivial, $\rho(C) = \max\{\rho(P) + 1 : P$ is a parent of $C\}$;
   (b) when C is trivial, $\rho(C) = \max\{\rho(P) + 1, \rho(Q) : P$ is a non-trivial parent of C, Q is a trivial parent of $C\}$.

In Definition 6.3, the first item indicates that if C has no parent, then C is located in a layer such that the status computation of arguments in C is independent of the status computation of any arguments in other SCCs.

The second item indicates that if C has some parents, then there are two possible cases:

- First, if C is non-trivial, then C is located in a layer such that the status of arguments in C is computed after the status of arguments in every parent of C has been computed.
- Second, if C is trivial, then C is located in the layer such that the status of arguments in C is computed after the status of arguments in every non-trivial parent[2] of C has been computed, and the status of arguments in every trivial parent of C, whose level is not higher than that of any non-trivial parent of C (when there exist some non-trivial parents of C), has been computed. In this way, the status of arguments in some trivial SCCs that have the same level can be computed together within a sub-framework, which, induced by the union of these trivial SCCs, is acyclic.

After all SCCs of an argumentation framework are assigned with specific levels, they can be easily decomposed into layers.

Definition 6.4 (Layer of SCCs). Let SCCS_F be the set of SCCs of an argumentation framework $F = (A, R)$, and $l_{max} = \max\{\rho(C) : C \in \text{SCCS}_F\}$.[3] Then, SCCS_F can be decomposed into $l_{max} + 1$ layers, denoted as $\text{SCCS}_F^0, \ldots,$ and $\text{SCCS}_F^{l_{max}}$, respectively, in which

$$\text{SCCS}_F^i = \{C \in \text{SCCS}_F \mid \rho(C) = i\} \tag{6.1}$$

Furthermore, let Trv_i and Non_i ($0 \le i \le l_{max}$) denote the set of trivial, and non-trivial SCCs of Layer i, respectively. Then, it holds that:

$$\text{SCCS}_F^i = Trv_i \cup Non_i \tag{6.2}$$

Example 6.3. According to Figure 6.2 and Definition 6.3, it holds that $\rho(C_1) = \rho(C_6) = \rho(C_7) = \rho(C_8) = 0$, $\rho(C_2) = \rho(C_3) = \rho(C_5) = 1$, and $\rho(C_4) = 2$. These SCCs are partitioned into three layers (as shown in Figure 6.3):

- In Layer 0, $\text{SCCS}_{F_{6.2}}^0 = \{C_1, C_6, C_7, C_8\}$, $Trv_0 = \{C_6, C_7, C_8\}$, and $Non_0 = \{C_1\}$.
- In Layer 1, $\text{SCCS}_{F_{6.2}}^1 = \{C_2, C_3, C_5\}$, $Trv_1 = \{C_2, C_3\}$, and $Non_1 = \{C_5\}$.
- In Layer 2, $\text{SCCS}_{F_{6.2}}^2 = \{C_4\}$, $Trv_2 = \emptyset$, and $Non_2 = \{C_4\}$.

In each layer i, $0 \le i \le l_{max}$, we use every non-trivial SCC to induce a sub-framework that contains cycles, and the union of all trivial SCCs to induce an acyclic sub-framework.

---

[2] Here, the term *trivial parent* (*non-trivial parent*) stands for a parent of a SCC that is trivial (respectively, non-trivial).

[3] Throughout this chapter, without explicit explanation, we use l_{max} to stand for $\max\{\rho(C) : C \in \text{SCCS}_F\}$, i.e., the highest level of a layered decomposition of an argumentation framework.

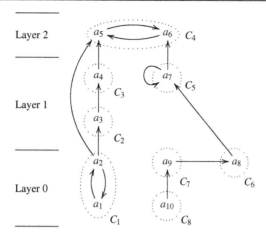

Figure 6.3 Layered decomposition of the strongly connected components of $F_{6.2}$.

Formally, let $U_i = \cup_{C_i \in Trv_i} C_i$ be the union of the set of trivial SCCs in Layer i. For every $P \in Non_i \cup \{U_i\}$, P is used to induce a sub-framework $(P \cup P^-, R_P \cup I_P)$.

The above formulation shows that given an argumentation framework, it can be decomposed into a set of sub-frameworks located in a number of layers.

Definition 6.5 (A decomposition of an argumentation framework).   Formally, a decomposition of an argumentation framework $F = (A, R)$, denoted as $\mathrm{decomp}(F)$, is defined as a tuple:

$$\mathrm{decomp}(F) = (\mathrm{decomp}_0(F), \dots, \mathrm{decomp}_{l_{max}}(F)) \tag{6.3}$$

In Formula (6.3), $\mathrm{decomp}_i(F) = \{(P \cup P^-, R_P \cup I_P) \mid P \in Non_i \cup \{U_i\}\}, 0 \le i \le l_{max}$.

Example 6.4.   Continue Example 6.3. According to Definition 6.5, $F_{6.2}$ is decomposed into a set of sub-frameworks located in three layers (Figure 6.4):

$\mathrm{decomp}(F_{6.2}) = (\mathrm{decomp}_0(F_{6.2}), \mathrm{decomp}_1(F_{6.2}), \mathrm{decomp}_2(F_{6.2}))$, where

- $\mathrm{decomp}_0(F_{6.2}) = \{(\{a_1, a_2\}, R_{\{a_1,a_2\}}), (\{a_8, a_9, a_{10}\}, R_{\{a_8,a_9,a_{10}\}})\}$.
- $\mathrm{decomp}_1(F_{6.2}) = \{(\{a_2, a_3, a_4\}, R_{\{a_3,a_4\}} \cup I_{\{a_3,a_4\}}), (\{a_7, a_8\}, R_{\{a_7\}} \cup I_{\{a_7\}})\}$.
- $\mathrm{decomp}_2(F_{6.2}) = \{(\{a_2, a_4, a_5, a_6, a_7\}, R_{\{a_5,a_6\}} \cup I_{\{a_5,a_6\}})\}$.

Now, let us verify the following dependence relations between different sub-frameworks in a decomposition. First, each sub-framework is not dependent on any other sub-frameworks in the same layer. Second, the sub-frameworks in a given layer are only dependent on the sub-frameworks in lower layers. Third, no two sub-frameworks are dependent on each other. Formally, we have the following proposition.

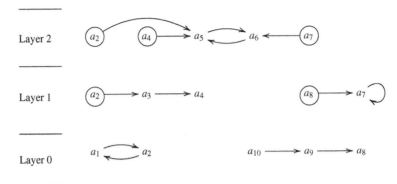

Figure 6.4 A decomposition of $F_{6.2}$.

Proposition 6.1.   *Let $F = (A, R)$ be an argumentation framework. Let $P \in Non_i \cup \{U_i\}$ and $Q \in Non_j \cup \{U_j\}$ be subsets of A, in which $0 \le i \le l_{max}, 0 \le j \le l_{max}$, and $P \ne Q$. In a decomposition of F (Definition 6.5), it holds that:*

(a) *If $i = j$, then $(P \cup P^-, R_P \cup I_P)$ is not dependent on $(Q \cup Q^-, R_Q \cup I_Q)$;*
(b) *If $(P \cup P^-, R_P \cup I_P)$ is dependent on $(Q \cup Q^-, R_Q \cup I_Q)$, then $i > j$.*
(c) *If $(P \cup P^-, R_P \cup I_P)$ is dependent on $(Q \cup Q^-, R_Q \cup I_Q)$, then it is not the case that $(Q \cup Q^-, R_Q \cup I_Q)$ is dependent on $(P \cup P^-, R_P \cup I_P)$.*

Proof.

(a) Assume that $(P \cup P^-, R_P \cup I_P)$ is dependent on $(Q \cup Q^-, R_Q \cup I_Q)$. It follows that there exist two SCCs $C_1 \subseteq P, C_2 \subseteq Q$, such that $\alpha \in C_1, \beta \in C_2$, and C_1 is a parent of C_2. Since P and Q are in the same layer, it holds that $\rho(C_1) = \rho(C_2)$. According to Definition 6.3, both C_1 and C_2 are trivial SCCs. It follows that C_1 and C_2 should belong to a unique sub-framework, and therefore $P = Q$. Contradiction.
(b) Since $(P \cup P^-, R_P \cup I_P)$ is dependent on $(Q \cup Q^-, R_Q \cup I_Q)$, according to Definition 4.2, there exist $\alpha \in P$ and $\beta \in Q$, such that there is a path from β to α with respect to R. So, there exist two SCCs $C_1 \subseteq P$ and $C_2 \subseteq Q$, such that $\alpha \in C_1, \beta \in C_2$, and $\rho(C_2) < \rho(C_1)$ (according to Definition 6.3). Hence, C_1 is in a layer higher than that of C_2, and as a result P is in a layer higher than that of Q. Since P is in Layer i and Q is in Layer j, it holds that $i > j$.
(c) Assume that $(Q \cup Q^-, R_Q \cup I_Q)$ is also dependent on $(P \cup P^-, R_P \cup I_P)$. It follows that there exist $\alpha, \beta \in P$ and $\gamma, \xi \in Q$, such that there are a path from α to γ and a path from ξ to β, with respect to R. Since there is a path from α to γ with respect to R, there exist two SCCs $C_1 \subseteq P$ and $C_2 \subseteq Q$, such that $\alpha \in C_1, \gamma \in C_2$, and C_1 is a parent of C_2. So, compared to Q, P is not located in a higher layer (so $j \ge i$). Meanwhile, since there is a path from ξ to β with respect to R, there exist two SCCs $C_1' \subseteq P$ and $C_2' \subseteq Q$,

such that $\xi \in C_2'$, $\beta \in C_1'$, and C_2' is a parent of C_1'. It follows that compared to P, Q is not located in a higher layer (so $j \le i$). As a result, P and Q are in the same layer ($j = i$). According to the proof of the item (a), $(P \cup P^-, R_P \cup I_P)$ is not dependent on $(Q \cup Q^-, R_Q \cup I_Q)$. Contradiction.

6.3 An Incremental Approach to Compute Argumentation Semantics

Given an argumentation framework, after it is decomposed into a set of sub-frameworks located in a number of layers, the computation of argumentation semantics proceeds incrementally, from the lowest layer (*Layer* 0) in which each sub-framework is not restricted by other sub-frameworks, to the highest layer (*Layer* l_{max}) in which each sub-framework is most restricted by the sub-frameworks located in the lower layers. In this chapter, we focus on the labelling-based approach.

6.3.1 The Computation of Layer i ($0 \le i \le l_{max}$)

The computation of a layer may include the following steps:

(1) Constructing partially labelled sub-frameworks in Layer i ($i \ge 1$),
(2) Computing the labellings of each sub-framework in Layer i ($i \ge 0$),
(3) Horizontally combining the labellings of Layer i ($i \ge 0$), and
(4) Vertically combining the labellings of layers from 0 to i ($i \ge 1$).

6.3.1.1 Constructing Partially Labelled Sub-Frameworks in Layer i ($i \ge 1$)

As presented in Section 4.3, given a sub-framework, if its set of conditioning arguments is not empty, then before computing the labellings of this sub-framework, we need to assign the status of the conditioning arguments externally and obtain a partially labelled sub-framework. Now, for all $(P \cup P^-, R_P \cup I_P) \in \text{decomp}_i(F)$, $1 \le i \le l_{max}$, P^- is not empty. So, the status of arguments in P^- could only be evaluated in the sub-frameworks that are not dependent on $(P \cup P^-, R_P \cup I_P)$.

According to Proposition 6.1, the sub-frameworks in a given layer are only dependent on the sub-frameworks in lower layers. So, given a sub-framework $(P \cup P^-, R_P \cup I_P) \in \text{decomp}_i(F)$, $1 \le i \le l_{max}$, the arguments in P^- are contained in a *single* sub-framework induced by the whole set of arguments located in layers from 0 to $i - 1$ (called *the multi-layer sub-framework, or briefly MLSF, of layers* from 0 to $i - 1$). Formally, we have the following definition and proposition.

Definition 6.6 (MLSF of layers from 0 to i). Let $F = (A, R)$ be an argumentation framework, and SCCS_F be a set of SCCs of F. The MLSF of layers from 0 to i, $0 \le i \le l_{max}$, is denoted as (M_i, R_{M_i}) (since M_i^- and I_{M_i} are empty, they are omitted), in which M_i is defined as follows:

$$M_i = \cup_{j=0}^{i} \left(\cup_{C \in SCCS_F^j} C \right) \tag{6.4}$$

Proposition 6.2.   *Based on Definitions 6.5 and 6.6, let $(P \cup P^-, R_P \cup I_P) \in decomp_i(F)$ be a sub-framework of $F = (A, R)$, $1 \leq i \leq l_{max}$. It holds that $P^- \subseteq M_{i-1}$.*

Proof.   According to Formula 2.1 in Chapter 2, $P^- = \{\alpha \in A \setminus P \mid \exists \beta \in P$, such that $(\alpha, \beta) \in R\}$. So, for every SCC $Q \subseteq A$, if $\exists \alpha \in Q \setminus P$, such that $(\alpha, \beta) \in R$, then there exists a SCC $C \subseteq P$, such that Q is a parent of C. According to Definition 6.3, $\rho(Q) \leq \rho(C) = i$. If $\rho(Q) = \rho(C)$, then Q and C are trivial SCCs (otherwise, $\rho(C) = \rho(Q) + 1$ and therefore $\rho(Q) \neq \rho(C)$). According to Definition 6.5, the union of all trivial SCCs in a layer is used to induce a unique sub-framework. It turns out that Q is included in P. So, $\nexists \alpha \in Q \setminus P$, such that $(\alpha, \beta) \in R$. Contradiction. As a result, $\rho(Q) \leq i - 1$, i.e., Q is located in a layer between 0 to $i - 1$. According to Formula (6.4) in Definition 6.6, all SCCs located in the layers from 0 to $i - 1$ are included in M_{i-1}. It follows that $P^- \subseteq M_{i-1}$.

Let $\mathcal{L}$ be a labeling of $(M_{i-1}, R_{M_{i-1}})$ under a given semantics σ.[4] Since $P^- \subseteq M_{i-1}$, according to Definition 4.3, $(P \cup P^-, R_P \cup I_P)^{\mathcal{L}}$ is a partially labelled sub-framework.

6.3.1.2 Computing the Labellings of Each Sub-Framework in Layer i (i ≥ 0)

In Layer 0, each sub-framework is independent. Its labellings are computed according to the definitions of various kinds of labellings presented in Section 2.3.2.

In Layer i, when $i \geq 1$, each sub-framework is dependent on the sub-frameworks in lower layers. As formulated in Section 6.3.1.1, before computing the semantics of each sub-framework in this layer, a set of partially labelled sub-frameworks are constructed according to the labellings of the MLSF of layers from 0 to $i - 1$. Then, for each partially labelled sub-framework, its labellings are computed according to Definition 4.6 in Chapter 4.

6.3.1.3 Horizontally Combining the Labellings of Layer i (i ≥ 0)

After the labellings of all sub-frameworks in a layer are computed, we hope that they could be combined to form the labellings of the sub-framework induced by the whole set of arguments in this layer. This process is called the *horizontal combination of argumentation semantics*. For convenience, the sub-framework induced by the whole set of arguments of a layer is called the *whole-layer sub-framework* (or briefly HLSF) of this layer. Formally, we have the following definition.

---

[4] In a layered computation approach, the labellings of an argumentation framework are computed incrementally from Layer 0 to Layer l_{max}. So, before computing the labellings of each sub-framework in Layer i, the (combined) labellings of the sub-framework induced by the MLSF of layers from 0 to $i - 1$ have been obtained.

Definition 6.7 (HLSF of Layer *i*). Let $F = (A, R)$ be an argumentation framework, and $SCCS_F$ be a set of SCCs of F. The HLSF of Layer i, $0 \leq i \leq l_{max}$, is denoted as $(H_i \cup H_i^-, R_{H_i} \cup I_{H_i})$, in which H_i is defined as follows:

$$H_i = \cup_{C \in SCCS_F^i} C \tag{6.5}$$

When $i = 0$, $(H_i \cup H_i^-, R_{H_i} \cup I_{H_i}) = (H_0, R_{H_0})$ is an independent sub-framework.

When $i \geq 1$, according to Proposition 6.2, it is easy to infer that H_i^- is a subset of M_{i-1}. Let $\mathcal{L}_\sigma((M_{i-1}, R_{M_{i-1}}))$ be the set of labellings of $(M_{i-1}, R_{M_{i-1}})$ under a given semantics σ. For every $\mathcal{L}$ in $\mathcal{L}_\sigma((M_{i-1}, R_{M_{i-1}}))$, it holds that $(H_i \cup H_i^-, R_{H_i} \cup I_{H_i})^{\mathcal{L}}$ is a partially labelled HLSF of Layer i.

Then, the labellings of an independent HLSF or a partially labelled HLSF are obtained by means of semantics combination.

Definition 6.8 (Combined labellings of Layer *i*). Let $decomp_i(F) = \{(P_1 \cup P_1^-, R_{P_1} \cup I_{P_1}), \ldots, (P_m \cup P_m^-, R_{P_m} \cup I_{P_m})\}$ be the set of sub-frameworks of $F = (A, R)$ within Layer i, in which $0 \leq i \leq l_{max}$ and $m \geq 1$. Let σ be a semantics.

First, the combined labellings of Layer 0, denoted as $CombLab_\sigma((H_0, R_{H_0}))$, is defined as follows:

$$CombLab_\sigma((H_0, R_{H_0})) = \{\mathcal{L}_1 + \ldots + \mathcal{L}_m \mid$$
$$\mathcal{L}_1 \in \mathcal{L}_\sigma((P_1, R_{P_1})) \wedge \ldots \wedge \mathcal{L}_m \in \mathcal{L}_\sigma((P_m, R_{P_m}))\}$$

Second, for every $\mathcal{L}$ in $\mathcal{L}_\sigma((M_{i-1}, R_{M_{i-1}}))$, $i \geq 1$, the combined labellings of Layer i, denoted as $CombLab_\sigma((H_i \cup H_i^-, R_{H_i} \cup I_{H_i})^{\mathcal{L}})$, is defined as follows:

$$CombLab_\sigma((H_i \cup H_i^-, R_{H_i} \cup I_{H_i})^{\mathcal{L}}) = \{\mathcal{L}_1 + \cdots + \mathcal{L}_m \mid$$
$$\mathcal{L}_1 \in \mathcal{L}_\sigma((P_1 \cup P_1^-, R_{P_1} \cup I_{P_1})^{\mathcal{L}}) \wedge \cdots \wedge$$
$$\mathcal{L}_m \in \mathcal{L}_\sigma((P_m \cup P_m^-, R_{P_m} \cup I_{P_m})^{\mathcal{L}})\}$$

6.3.1.4 Vertically Combining the Labellings of Layers from 0 to i (i ≥ 1)

After we get the labellings of the HLSF of Layer i ($1 \leq i \leq l_{max}$) and the labellings of the MLSF of layers from 0 to $i - 1$ (initially, when $i = 1$, the MLSF of layers from 0 to $i - 1$ is equal to the HLSF of Layer 0), we hope that they could be combined to form the labellings of the MLSF of layers from 0 to i. This process is called a *vertical combination* of argumentation semantics of layers from 0 to i. Formally, we have the following definition.

Definition 6.9 (Combined labellings of layers from 0 to *i*). Let σ be a semantics. Let $F = (A, R)$ be an argumentation framework, $(M_{i-1}, R_{M_{i-1}})$ be the MLSF of layers from 0 to

$i - 1$, and $(H_i \cup H_i^-, R_{H_i} \cup I_{H_i})$ be the HLSF of Layer i, $1 \le i \le l_{max}$. The set of *combined labellings* of (M_i, R_{M_i}), denoted as $CombLab_\sigma((M_i, R_{M_i}))$, is defined as follows:

$$CombLab_\sigma((M_i, R_{M_i})) = \{\mathcal{L}_1 + \mathcal{L}_2 \mid \mathcal{L}_1 \in \mathcal{L}_\sigma((M_{i-1}, R_{M_{i-1}})) \wedge$$
$$\mathcal{L}_2 \in \mathcal{L}_\sigma((H_i \cup H_i^-, R_{H_i} \cup I_{H_i})^{\mathcal{L}_1})\}$$

After introducing the incremental approach for computing argumentation semantics, it is necessary to verify that this approach is correct. In the next subsection, we will explain the soundness and completeness of semantic combination, which ensure the correctness of the incremental approach. Then, in Section 6.3.3, we will further explain the process and the correctness of the incremental approach by an example.

6.3.2 Soundness and Completeness of Semantic Combination

Whether the combined labellings are the labellings under a given semantics depends on the soundness and completeness of the semantic combination in all layers. According to theories presented in Chapter 5, we directly have the following propositions.

The first proposition is for the vertical semantics combination.

Proposition 6.3. *Let $\sigma \in \{adm, co, pr, gr\}$ be a semantics. Let $F = (A, R)$ be an argumentation framework, $(M_{i-1}, R_{M_{i-1}})$ be the MLSF of layers from 0 to $i - 1$, and $(H_i \cup H_i^-, R_{H_i} \cup I_{H_i})$ be the HLSF of Layer i, $1 \le i \le l_{max}$. The vertical combination of argumentation semantics of layers from 0 to i is sound and complete. Formally, it holds that:*

$$\mathcal{L}_\sigma((M_i, R_{M_i})) = CombLab_\sigma((M_i, R_{M_i}))$$

The second is for the horizontal semantic combination.

Proposition 6.4. *Let $\sigma \in \{adm, co, pr, gr\}$ be a semantics. Let $F = (A, R)$ be an argumentation framework. Let $(H_i \cup H_i^-, R_{H_i} \cup I_{H_i})$ be the HLSF of Layer i, $0 \le i \le l_{max}$. Let $(M_{i-1}, R_{M_{i-1}})$ be the MLSF of layers from 0 to $i - 1$, $1 \le i \le l_{max}$. The horizontal combination of argumentation semantics is sound and complete. Specifically, it holds that:*

(1) $\mathcal{L}_\sigma((H_0, R_{H_0})) = CombLab_\sigma((H_0, R_{H_0}))$; and

(2) for all $\mathcal{L} \in \mathcal{L}_\sigma((M_{i-1}, R_{M_{i-1}}))$, $1 \le i \le l_{max}$, it holds that $\mathcal{L}_\sigma((H_i \cup H_i^-, R_{H_i} \cup I_{H_i})^{\mathcal{L}}) = CombLab_\sigma((H_i \cup H_i^-, R_{H_i} \cup I_{H_i})^{\mathcal{L}})$.

6.3.3 An Illustrating Example

In order to further illustrate the process of the incremental approach, let us consider the following example. In this example, we only consider the computation under preferred semantics.

Example 6.5. According to the decomposition of $F_{6.2}$ in Example 6.4, the computation proceeds from Layer 0 to Layer 2.

In Layer 0, there are the following two steps of computation.

Step 0_1: Computing the preferred labellings of each independent sub-framework. According to Example 6.4, we have $decomp_0(F_{6.2}) = \{(\{a_1, a_2\}, R_{\{a_1,a_2\}}), (\{a_8, a_9, a_{10}\}, R_{\{a_8,a_9,a_{10}\}})\}$. According to Definition 2.18 in Chapter 2, it holds that:

$$\mathscr{L}_{pr}((\{a_1, a_2\}, R_{\{a_1,a_2\}})) = \{\mathfrak{L}_1, \mathfrak{L}_2\}$$
$$\mathscr{L}_{pr}((\{a_8, a_9, a_{10}\}, R_{\{a_8,a_9,a_{10}\}})) = \{\mathfrak{L}_3\}$$
$$\mathfrak{L}_1 = (\{a_1\}, \{a_2\}, \emptyset)$$
$$\mathfrak{L}_2 = (\{a_2\}, \{a_1\}, \emptyset)$$
$$\mathfrak{L}_3 = (\{a_8, a_{10}\}, \{a_9\}, \emptyset)$$

Step 0_2: Horizontally combining the labellings of Layer 0. By combining the set of preferred labellings of $(\{a_1, a_2\}, R_{\{a_1,a_2\}})$ and the set of preferred labellings of $(\{a_8, a_9, a_{10}\}, R_{\{a_8,a_9,a_{10}\}})$, we obtain the set of combined labellings of $(H_0, R_{H_0}) = (\{a_1, a_2, a_8, a_9, a_{10}\}, \{(a_1, a_2), (a_2, a_1), (a_{10}, a_9), (a_9, a_8)\})$, the HLSF of Layer 0:

$$CombLab_{pr}((H_0, R_{H_0})) = \{\mathfrak{L}_4, \mathfrak{L}_5\}$$
$$\mathfrak{L}_4 = \mathfrak{L}_1 + \mathfrak{L}_3 = (\{a_1, a_8, a_{10}\}, \{a_2, a_9\}, \emptyset)$$
$$\mathfrak{L}_5 = \mathfrak{L}_2 + \mathfrak{L}_3 = (\{a_2, a_8, a_{10}\}, \{a_1, a_9\}, \emptyset)$$

According to Proposition 6.4, $\mathscr{L}_{pr}((H_0, R_{H_0})) = \{\mathfrak{L}_4, \mathfrak{L}_5\}$.

In Layer 1, there are the following four steps of computation.

Step 1_1: Constructing partially labelled sub-frameworks in Layer 1. According to Example 6.4, $decomp_1(F_{6.2}) = \{((\{a_2, a_3, a_4\}, R_{\{a_3,a_4\}} \cup I_{\{a_3,a_4\}}), (\{a_7, a_8\}, R_{\{a_7\}} \cup I_{\{a_7\}})\}$. When $i = 1$, $(M_0, R_{M_0}) = (H_0, R_{H_0})$ (the MLSF of layers from 0 to $i - 1$ is equal to the HLSF of Layer 0). Since (M_0, R_{M_0}) has two labellings, we obtain four partially labelled sub-frameworks as follows (Figure 6.5):

$$(\{a_2, a_3, a_4\}, R_{\{a_3,a_4\}} \cup I_{\{a_3,a_4\}})^{\mathfrak{L}_4}$$
$$(\{a_2, a_3, a_4\}, R_{\{a_3,a_4\}} \cup I_{\{a_3,a_4\}})^{\mathfrak{L}_5}$$
$$(\{a_7, a_8\}, R_{\{a_7\}} \cup I_{\{a_7\}})^{\mathfrak{L}_4}$$
$$(\{a_7, a_8\}, R_{\{a_7\}} \cup I_{\{a_7\}})^{\mathfrak{L}_5}$$

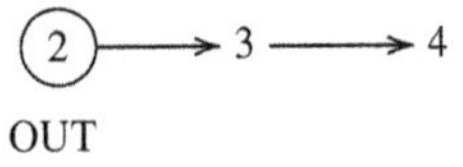

(a) $(\{a_2, a_3, a_4\}, R_{\{a_3,a_4\}} \cup I_{\{a_3,a_4\}})^{\mathcal{L}_4}$

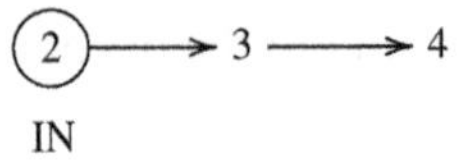

(b) $(\{a_2, a_3, a_4\}, R_{\{a_3,a_4\}} \cup I_{\{a_3,a_4\}})^{\mathcal{L}_5}$

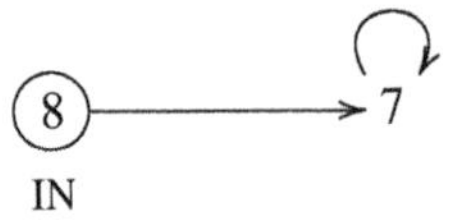

(c) $(\{a_7, a_8\}, R_{\{a_7\}} \cup I_{\{a_7\}})^{\mathcal{L}_4}$; $(\{a_7, a_8\}, R_{\{a_7\}} \cup I_{\{a_7\}})^{\mathcal{L}_5}$

Figure 6.5   Four partially labelled sub-frameworks of $F_{6.2}$ within Layer 1.

Step 1_2: Computing the labellings of each partially labelled sub-framework in Layer 1. According to Definition 4.6 in Chapter 4, it holds that:

$$\mathscr{L}_{pr}((\{a_2, a_3, a_4\}, R_{\{a_3,a_4\}} \cup I_{\{a_3,a_4\}})^{\mathcal{L}_4}) = \{\mathcal{L}_6\}$$
$$\mathscr{L}_{pr}((\{a_2, a_3, a_4\}, R_{\{a_3,a_4\}} \cup I_{\{a_3,a_4\}})^{\mathcal{L}_5}) = \{\mathcal{L}_7\}$$
$$\mathscr{L}_{pr}((\{a_7, a_8\}, R_{\{a_7\}} \cup I_{\{a_7\}})^{\mathcal{L}_4}) = \{\mathcal{L}_8\}$$
$$\mathscr{L}_{pr}((\{a_7, a_8\}, R_{\{a_7\}} \cup I_{\{a_7\}})^{\mathcal{L}_5}) = \{\mathcal{L}_9\}$$
$$\mathcal{L}_6 = (\{a_3\}, \{a_2, a_4\}, \emptyset)$$
$$\mathcal{L}_7 = (\{a_2, a_4\}, \{a_3\}, \emptyset)$$
$$\mathcal{L}_8 = (\{a_8\}, \{a_7\}, \emptyset)$$
$$\mathcal{L}_9 = (\{a_8\}, \{a_7\}, \emptyset)$$

Step 1_3: Horizontally combining the semantics of Layer 1. The HLSF of Layer 1 is $(H_1 \cup H_1^-, R_{H_1} \cup I_{H_1}) = (\{a_2, a_3, a_4, a_7, a_8\}, \{(a_2, a_3), (a_3, a_4), (a_7, a_7), (a_8, a_7)\})$, in which the set of conditioning arguments is $H_1^- = \{a_2, a_8\}$. The corresponding two partially labelled sub-frameworks are $(H_1 \cup H_1^-, R_{H_1} \cup I_{H_1})^{\mathcal{L}_4}$ and $(H_1 \cup H_1^-, R_{H_1} \cup I_{H_1})^{\mathcal{L}_5}$ (Figure 6.6). According to Definition 6.8, it holds that:

$$CombLab_{pr}((H_1 \cup H_1^-, R_{H_1} \cup I_{H_1})^{\mathcal{L}_4}) = \{\mathcal{L}_{10}\}$$
$$CombLab_{pr}((H_1 \cup H_1^-, R_{H_1} \cup I_{H_1})^{\mathcal{L}_5}) = \{\mathcal{L}_{11}\}$$
$$\mathcal{L}_{10} = \mathcal{L}_6 + \mathcal{L}_8 = (\{a_3, a_8\}, \{a_2, a_4, a_7\}, \emptyset)$$
$$\mathcal{L}_{11} = \mathcal{L}_7 + \mathcal{L}_9 = (\{a_2, a_4, a_8\}, \{a_3, a_7\}, \emptyset)$$

According to Proposition 6.4, we have $\mathscr{L}_{pr}((H_1 \cup H_1^-, R_{H_1} \cup I_{H_1})^{\mathcal{L}_4}) = \{\mathcal{L}_{10}\}$, and $\mathscr{L}_{pr}((H_1 \cup H_1^-, R_{H_1} \cup I_{H_1})^{\mathcal{L}_5}) = \{\mathcal{L}_{11}\}$.

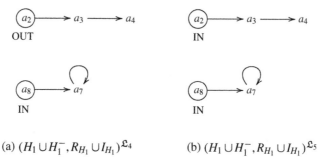

(a) $(H_1 \cup H_1^-, R_{H_1} \cup I_{H_1})^{\mathcal{L}_4}$ (b) $(H_1 \cup H_1^-, R_{H_1} \cup I_{H_1})^{\mathcal{L}_5}$

Figure 6.6   Two partially labelled HLSFs of $F_{6.2}$ within Layer 1.

Step 1_4: Vertically combining the semantics of layers from 0 to 1. According to Example 6.3 and Formula 6.4 in Definition 6.6, $M_1 = \{a_1, a_2, a_3, a_4, a_7, a_8, a_9, a_{10}\}$. So, the MLSF of layers from 0 to 1 is (M_1, R_{M_1}). According to Definition 6.9, it holds that:

$$CombLab_{pr}((M_1, R_{M_1})) = \{\mathcal{L}_{12}, \mathcal{L}_{13}\}$$
$$\mathcal{L}_{12} = \mathcal{L}_4 + \mathcal{L}_{10} = (\{a_1, a_3, a_8, a_{10}\}, \{a_2, a_4, a_7, a_9\}, \emptyset)$$
$$\mathcal{L}_{13} = \mathcal{L}_5 + \mathcal{L}_{11} = (\{a_2, a_4, a_8, a_{10}\}, \{a_1, a_3, a_7, a_9\}, \emptyset)$$

According to Proposition 6.3, $\mathscr{L}_{pr}((M_1, R_{M_1})) = \{\mathcal{L}_{12}, \mathcal{L}_{13}\}$.

Finally, in Layer 2, there are also four steps of computation.

Step 2_1: Constructing partially labelled sub-frameworks in Layer 2. According to Example 6.4, $decomp_2(F_{6.2}) = \{(\{a_2, a_4, a_5, a_6, a_7\}, R_{\{a_5, a_6\}} \cup I_{\{a_5, a_6\}})\}$. Since (M_1, R_{M_1}) has two labellings, we obtain two partially labelled sub-frameworks as follows (Figure 6.7).

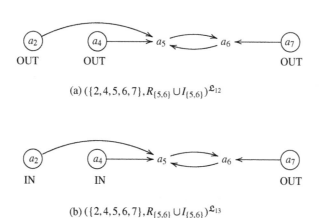

(a) $(\{2, 4, 5, 6, 7\}, R_{\{5,6\}} \cup I_{\{5,6\}})^{\mathcal{L}_{12}}$

(b) $(\{2, 4, 5, 6, 7\}, R_{\{5,6\}} \cup I_{\{5,6\}})^{\mathcal{L}_{13}}$

Figure 6.7   Two partially labelled sub-frameworks of $F_{6.2}$ within Layer 2.

Step 2_2: Computing the semantics of each partially labelled sub-framework in Layer 2. According to Definition 4.6 in Chapter 4, it holds that:

$$\mathscr{L}_{pr}((\{a_2, a_4, a_5, a_6, a_7\}, R_{\{a_5,a_6\}} \cup I_{\{a_5,a_6\}})^{\mathscr{L}_{12}}) = \{\mathscr{L}_{14}, \mathscr{L}_{15}\}$$
$$\mathscr{L}_{pr}((\{a_2, a_4, a_5, a_6, a_7\}, R_{\{a_5,a_6\}} \cup I_{\{a_5,a_6\}})^{\mathscr{L}_{13}}) = \{\mathscr{L}_{16}\}$$

$$\mathscr{L}_{14} = (\{a_5\}, \{a_2, a_4, a_6, a_7\}, \emptyset)$$
$$\mathscr{L}_{15} = (\{a_6\}, \{a_2, a_4, a_5, a_7\}, \emptyset)$$
$$\mathscr{L}_{16} = (\{a_2, a_4, a_6\}, \{a_5, a_7\}, \emptyset)$$

Step 2_3: Horizontally combining the semantics of Layer 2. Since there is only one sub-framework in this layer, the semantics combination is trivial. Let $(H_2 \cup H_2^-, R_{H_2} \cup I_{H_2})$ be the HLSF of Layer 2. $(H_2 \cup H_2^-, R_{H_2} \cup I_{H_2}) = (\{a_2, a_4, a_5, a_6, a_7\}, \{(a_2, a_5), (a_4, a_5), (a_5, a_6), (a_6, a_5), (a_7, a_6)\})$. It is obvious that

$$\mathscr{L}_{pr}((H_2 \cup H_2^-, R_{H_2} \cup I_{H_2})^{\mathscr{L}_{12}}) = \{\mathscr{L}_{14}, \mathscr{L}_{15}\}$$
$$\mathscr{L}_{pr}((H_2 \cup H_2^-, R_{H_2} \cup I_{H_2})^{\mathscr{L}_{13}}) = \{\mathscr{L}_{16}\}$$

Step 2_4: Vertically combining the semantics of layers from 0 to 2. The MLSF of layers from 0 to 2 is $(M_2, R_{M_2}) = (A, R) = F_{6.2}$. According to Definition 6.9, it holds that:

$$CombLab_{pr}((M_2, R_{M_2})) = \{\mathscr{L}_{17}, \mathscr{L}_{18}, \mathscr{L}_{19}\}$$

$$\mathscr{L}_{17} = \mathscr{L}_{12} + \mathscr{L}_{14} = (\{a_1, a_3, a_5, a_8, a_{10}\}, \{a_2, a_4, a_6, a_7, a_9\}, \emptyset)$$
$$\mathscr{L}_{18} = \mathscr{L}_{12} + \mathscr{L}_{15} = (\{a_1, a_3, a_6, a_8, a_{10}\}, \{a_2, a_4, a_5, a_7, a_9\}, \emptyset)$$
$$\mathscr{L}_{19} = \mathscr{L}_{13} + \mathscr{L}_{16} = (\{a_2, a_4, a_6, a_8, a_{10}\}, \{a_1, a_3, a_5, a_7, a_9\}, \emptyset)$$

Since $(M_2, R_{M_2}) = (A, R) = F_{6.2}$, the set of combined extensions of $F_{6.2}$ is $\{\mathscr{L}_{17}, \mathscr{L}_{18}, \mathscr{L}_{19}\}$. According to Proposition 6.3, $\mathscr{L}_{pr}(F_{6.2}) = \{\mathscr{L}_{17}, \mathscr{L}_{18}, \mathscr{L}_{19}\}$.

6.4 Empirical Evaluation

In this section, for simplicity and without loss of generality, we only focus on evaluating the performance of our decomposition-based algorithm for computing *preferred* labellings of an argumentation framework. The algorithm is composed of the following two components.

The first component is responsible for decomposing argumentation frameworks. Given an argumentation framework $F = (A, R)$, it is partitioned into a set of SCCs (by using Tarjan's algorithm [14]). Then, each SCC is assigned with a *level* (according to Definition 6.3), decomposition of F is obtained.

The second component is responsible for the incremental computation of preferred labellings. Besides the combination of labellings (as formulated in Section 5.3), the core of this component is to compute the preferred labellings of each sub-framework. Given an independent sub-framework (P, R_P) of F, its preferred labellings are computed directly by the MC algorithm; given a partially labelled sub-framework $(P \cup P^-, R_P \cup I_P)^{\mathcal{L}}$ of F, in which $P^- \neq \emptyset$, its preferred labellings are computed by Algorithm 4.1 in Chapter 4.

The algorithm was implemented in Java, and tested on a machine with an Intel CPU running at 1.86 GHz and 1.98 GB RAM. Since the performance of the algorithms for computing the labellings of an argumentation framework is highly related to its topology and edge density, we tried the following three configurations:

- The nodes of argumentation frameworks range from 100 to 200, and the ratio of the number of edges to the number of nodes is 1:1.
- The nodes of argumentation frameworks range from 15 to 25, and the ratios of the number of edges to the number of nodes are respectively 1.5:1 and 2:1.
- The nodes of argumentation frameworks range from 8 to 12, and the ratios of the number of edges to the number of nodes are respectively 3:1, 4:1 and 5:1.

In the test, for each assignment of the number of nodes and the ratio of the number of edges to the number of nodes, the program was executed 20 times. In each time, an argumentation framework was generated at random,[5] and then its preferred labellings were computed by using the MC algorithm and our decomposition-based algorithm (or briefly, our algorithm), respectively. Table 6.1 shows the average results of the number of SCCs and the execution time of the two approaches.

In this table, "#nodes", "#SCCS" and "#timeout"denote, respectively, the number of nodes of an argumentation framework, the number of SCCs, and the number of timeouts (when the time for computing the preferred labellings of an argumentation framework is over 30 minutes) among the 20 times of execution.[6] When the average time shown in Table 6.1 was computed, only the cases without timeouts were considered. For instance, when #nodes = 180 and the ratio of #edges to #nodes is 1:1, there are four timeouts. According to the corresponding records in Table 6.3, the average time of the MC algorithm is: $(0.015 + 0.015 + 0.016 + 0.015 + 0.015 + 0.015 + 0.016 + 0.016 + 22.781 + 0.016 + 3.485 + 0.032 + 0.016 + 0.015 + 0.016 + 0.016) \div 16 = 1.656$; the average time of our algorithm is:

---

[5] The basic idea of generating a random graph is as follows: Given a set of arguments $\{1, 2, \ldots, n\}$ and a ratio a, select at random two nodes x_1 and x_2 from the set, if the edge (x_1, x_2) has not been generated, then (x_1, x_2) is generated, until the number of edges is equal to a given value $(n \times a)$.

[6] Since in many cases, the execution time might last very long, to make the test easier, when the time for computing the preferred labellings of an argumentation framework is over 30 minutes, we stopped the execution by setting a break in the program.

Table 6.1 Average results of the two algorithms.

Ratio	#nodes	#SCCS	Our Alg.		MC Alg.	
			Time (seconds)	#timeout	Time (seconds)	#timeout
	100	95	0.013	0	1.444	0
	120	114	0.008	0	0.022	3
1:1	140	135	0.012	0	0.055	3
	160	156	0.015	0	5.033	1
	180	**171**	**0.017**	**0**	**1.656**	**4**
	200	192	0.019	0	0.328	1
	15	**10**	**0.451**	**0**	**9.818**	**0**
	17	11	0.015	0	54.236	2
1.5:1	19	11	0.001	0	57.379	2
	21	12	0.002	1	1.775	3
	23	16	4.496	0	40.126	2
	25	18	0.002	0	2.933	3
	15	7	2.133	0	95.911	4
	17	7	1.501	1	12.971	8
2:1	19	9	0	3	0.576	5
	21	10	0.001	1	0.59	6
	23	9	0.451	3	24.94	7
	25	10	0.119	3	34.87	4
	8	2	0.026	0	0.050	0
	9	3	0.122	0	0.153	0
3:1	10	3	0.342	0	0.970	0
	11	3	1.685	0	8.185	0
	12	3	11.510	0	72.841	0
	8	2	0.064	0	0.081	0
	9	2	0.382	0	0.538	0
4:1	10	2	2.839	0	3.880	0
	11	2	21.257	0	25.793	0
	12	2	166.641	0	234.396	0
	8	2	0.116	0	0.130	0
	9	2	0.856	0	0.953	0
5:1	10	2	5.622	0	6.409	0
	11	2	41.482	0	50.497	0
	12	2	449.471	0	497.104	0

$(0.016 + 0.031 + 0.016 + 0.016 + 0.016 + 0.016 + 0.016 + 0.016 + 0.016 + 0.015 + 0.031 + 0.016 + 0.015 + 0.015 + 0.015 + 0.015 + 0.016 + 0.016 + 0.015 + 0.015) \div 20 = 0.017$.

Note that in our decomposition-based algorithm, the execution time includes the time for generating a set of SCCs, for constructing a set of sub-frameworks, and for generating and combining the preferred labellings of all sub-frameworks. Compared to the time for generating preferred labellings, the time for computing a set of SCCs is negligible. Table 6.2

Table 6.2 The average time for computing SCCs (in milliseconds).

#nodes	Ratio = 1:1	Ratio = 1.5:1	Ratio = 2:1	Ratio = 3:1	Ratio = 4:1	Ratio = 5:1
10	0.15	0.16	0.31	0	0.16	0
30	0	0.15	0.16	0	0.16	0
50	0.16	0.16	0.31	0	0	0.32
70	0.32	0.31	0	0.31	0	0.15
90	0	0.15	0.46	0.15	0.62	0.46
110	0.61	0.46	0.63	0.32	0.79	0.78
130	0.30	0	0.63	0.47	0.94	0.94
150	0.78	0.62	0.93	0.93	1.25	1.09
170	0.31	0.95	1.24	0.48	1.08	1.71
190	0.61	0.31	0.76	0.93	1.72	1.25

shows that the average time for computing the SCCs of argumentation frameworks in all cases is less than 2 milliseconds.[7]

From Table 6.1, we found that when the ratio of the number of edges to the number of nodes is around 1.5:1, our decomposition-based algorithm is very efficient. With the increase of the ratios of the number of edges to the number of nodes, the ratios of the number of SCCs to the number of nodes decrease. Meanwhile, the execution time of the two algorithms increases, and the computational advantage of our algorithm becomes more and more indistinct.

Now, let us have a close look at the detailed results of the individual cases of computation. Table 6.3 shows the cases when #nodes = 180 and the ratio of the number of edges to the number of nodes is 1:1. In this table, #undec denotes the maximal number of arguments that are labelled UNDEC in a preferred labelling of a given argumentation framework; #lab denotes the number of preferred labellings of a given argumentation framework; "Tri[..., $i(n_i)$, ...,]" indicates the number of arguments in trivial SCCs of Layer i; "Non[..., $j(n_{j1} + \cdots + n_{jm})$, ...]" indicates the number of non-trivial SCCs in Layer j and the number of arguments in each non-trivial SCC. For instance, in Record 1 of Table 6.3, "Tri[0(111), 1(2), 2(25), 4(16)]" denotes that the number of arguments in trivial SCCs of Layers 0, 1, 2 and 4 is respectively 111, 2, 25 and 16; "Non[0(1+1), 1(15), 3(9)]" denotes that there are two non-trivial SCCs (each of which contains one argument) in Layer 0, one non-trivial SCC (which contains 15 arguments) in Layer 1, and one non-trivial SCC (which contains 9 arguments) in Layer 3. For simplicity, when there are no trivial (or non-trivial) SCCs in a given layer, the corresponding item is omitted.

---

[7] The data of Table 6.2 were generated in another test. Given an assignment of edge density (ration = 1:1, 1.5:1, 2:1, 3:1, 4:1, 5:1) and the number of nodes (10, 30, ..., 190), the program for computing SCCs was executed 100 times. In each time, an argumentation framework with given edge density and size was generated at random, and the set of SCCs of the argumentation framework was computed.

Table 6.3   The cases when #nodes = 180 and the ratio of #edges to #nodes is 1 : 1.

No.	#SCC	Our Alg. (seconds)	MC Alg. (seconds)	#undec	#lab	Details of each decomposition
1	158	0.016	0.015	3	1	Tri[0(111), 1(2), 2(25), 4(16)], Non[0(1+1), 1(15), 3(9)]
2	158	0.031	timeout	21	1	Tri[0(135), 2(33)], Non[1(1+2+2+7)]
3	177	0.016	0.015	0	1	Tri[0(162), 2(14)], Non[1(4)]
4	177	0.016	timeout	12	2	Tri[0(157), 1(1), 2(15)], Non[0(1), 1(1+4+1)]
5	180	0.016	0.016	0	1	Tri[0(179)], Non[1(1)]
6	180	0.016	0.015	0	1	Tri[0(157), 2(22)], Non[1(1)]
7	176	0.016	0.015	0	1	Tri[0(160), 2(15)], Non[1(5)]
8	150	0.016	0.015	0	1	Tri[0(121), 2(28)], Non[1(31)]
9	176	0.016	0.016	0	1	Tri[0(133), 2(25), 4(14)], Non[1(3+1+2), 3(2)]
10	176	0.015	0.016	0	1	Tri[0(161), 2(14)], Non[1(5)]
11	173	0.031	**22.781**	0	2	Tri[0(157), 2(14)], Non[1(7+2)]
12	170	0.016	0.016	0	1	Tri[0(128), 2(14), 4(25)], Non[1(1), 3(9+3)]
13	178	0.015	**3.485**	9	1	Tri[0(171), 2(6)], Non[1(3)]
14	174	0.015	0.032	5	1	Tri[0(142), 1(7), 2(21)], Non[0(1), 1(2+1+6)]
15	144	0.015	timeout	4	1	Tri[0(98), 1(3), 2(4), 4(35)], Non[0(1+1), 1(1), 3(37)]
16	178	0.015	0.016	1	1	Tri[0(169), 1(4), 2(3)], Non[0(1), 1(3)]
17	170	0.016	0.015	0	1	Tri[0(122), 2(16), 4(21), 6(8)], Non[1(7), 3(4), 5(2)]
18	177	0.016	timeout	8	2	Tri[0(125), 2(50)], Non[1(4+1)]
19	171	0.015	0.016	2	1	Tri[0(160), 2(8)], Non[0(1), 1(1+10)]
20	179	0.015	0.016	3	1	Tri[0(173), 1(2), 2(2)], Non[0(1), 1(2)]
avg.	**171**	**0.017**	**1.656**			

From Table 6.3, we found that the execution time of our decomposition-based algorithm is very low (no more than 0.031 seconds) in all cases, while the execution time of the MC algorithm fluctuates from 0.015 seconds to more than 30 minutes. What are the basic reasons behind this phenomenon? Intuitively, we may observe that the time complexity of the MC algorithm is closely related to the number of arguments that are labelled UNDEC and the number of preferred labellings. Please look at Records 2, 4, 11, 13, 15 and 18 in Table 6.3: in Record 2, #undec = 21 and the execution time is more than 30 minutes; in Record 13, #undec = 9 and the execution time is 3.485 seconds; in Record 15, #undec = 4 and the execution time is more than 30 minutes; in Records 4, 11 and 18, all argumentation frameworks have two preferred labellings. Theoretically, according to the MC algorithm for generating preferred labellings, when the number of UNDEC-labelled arguments is bigger, it takes more time to perform transition steps. Meanwhile, when an argumentation framework

Table 6.4 The cases when #nodes = 15 and the ratio of #edges to #nodes is 1.5:1.

No.	#SCC	Our Alg. (seconds)	MC Alg. (seconds)	#undec	#lab	Details of each decomposition
1	4	9.016	10.062	8	1	**Tri[0(2), 2(1)], Non[1(12)]**
2	11	0	0.75	6	2	Tri[0(1), 2(6), 3(1)], Non[0(2), 1(3), 2(2)]
3	7	0	0	0	1	Tri[0(5), 2(1)], Non[1(9)]
4	11	0	0.047	0	3	Tri[0(3), 2(7)], Non[1(5)]
5	8	0	0	0	1	Tri[0(7)], Non[1(8)]
6	15	0	2.484	10	1	**Tri[0(5), 2(9)], Non[1(1)]**
7	9	0	0	0	2	Tri[0(5), 2(3)], Non[1(7)]
8	14	0	0	3	1	Tri[0(4), 2(6)], Non[0(1+1), 1(1+2)]
9	12	0	0	1	1	Tri[0(4), 2(6)], Non[1(4), 2(1)]
10	11	0	0	0	1	Tri[0(6), 2(4)], Non[1(5)]
11	12	0	0	1	1	Tri[0(3), 2(6)], Non[1(1+4), 3(1)]
12	11	0	0	1	1	Tri[0(8)], Non[1(1+1), 2(5)]
13	14	0	0	0	1	Tri[0(7), 2(1), 3(4)], Non[1(2), 2(1)]
14	7	0.015	183.031	3	3	**Tri[0(2), 3(1)], Non[0(4), 1(2), 2(5+1)]**
15	14	0	0	0	2	Tri[0(8), 2(3)], Non[1(1+2+1)]
16	9	0	0	0	1	Tri[0(8)], Non[1(7)]
17	10	0	0	1	1	Tri[0(5), 2(3)], Non[0(1), 1(6)]
18	9	0	0	0	1	Tri[0(5), 1(2)], Non[0(2), 2(6)]
19	13	0	0	4	1	Tri[0(8), 3(1)], Non[0(1), 1(1), 2(1+3)]
20	10	0	0	6	1	Tri[0(1), 2(6)], Non[0(1), 1(6), 2(1)]
avg.	**10**	**0.451**	**9.818**			

has more preferred labellings, the time for finding and comparing candidate labellings tends to increase.

To see the specific topologies of the argumentation frameworks that have more UNDEC-labelled arguments and/or more preferred labellings, let us consider the cases when #nodes = 15 and the ratio of the number of edges to the number of nodes is 1.5:1 (Table 6.4).

According to Record 1 in Table 6.4, the number of UNDEC-labelled arguments is 8, and the execution times of our algorithm and MC algorithm are, respectively, 9.016 seconds and 10.062 seconds. The corresponding argumentation framework is illustrated in Figure 6.8(a). This argumentation framework has only one preferred labelling ($\{a_4, a_6, a_{12}\}$, $\{a_1, a_3, a_5, a_{11}\}$, $\{a_2, a_7, a_8, a_9, a_{10}, a_{13}, a_{14}, a_{15}\}$). According to the topology of this argumentation framework, we found that the UNDEC-labelled arguments are related to some odd-length cycles. In this example, there are three odd-length cycles ($\{a_5\}$, $\{(a_5, a_5)\}$), ($\{a_{10}\}$, $\{(a_{10}, a_{10})\}$) and ($\{a_7, a_8, a_9, a_{10}, a_{13}\}$, $\{(a_7, a_9), (a_9, a_{10}), (a_{10}, a_8), (a_8, a_{13}), (a_{13}, a_9)\}$), which are the fundamental reason that the related arguments are undecided under preferred semantics. Meanwhile, we observed that there is a non-trivial SCC whose size is 12 in this argumentation framework. This is why the execution time of our algorithm is close to that of

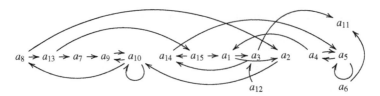

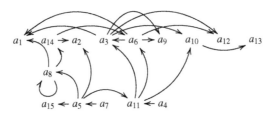

(a) The argumentation framework with respect to Record 1

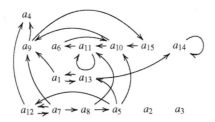

(b) The argumentation framework with respect to Record 6

(c) The argumentation framework with respect to Record 14

Figure 6.8   Argumentation frameworks with respect to Records 1, 6 and 14 in Table 6.4.

the MC algorithm. In other words, the efficiency of our algorithm is highly limited by the size of the maximal SCC of an argumentation framework.

The argumentation framework with respect to Record 6, as shown in Figure 6.8(b), is interesting. It has only one non-trivial SCC, which is an odd-length cycle containing only one argument. However, from the unique preferred labelling ($\{a_4, a_7, a_{15}\}$, $\{a_5, a_{11}\}$, $\{a_1, a_2, a_3, a_6, a_8, a_9, a_{10}, a_{12}, a_{13}, a_{14}\}$), we observed that when the arguments in acyclic fragments are affected by some arguments from odd-length cycles, they might also become undecided under preferred semantics and consume more computation time.

Finally, as illustrated in Figure 6.8(c), the argumentation framework with respect to Record 14 has four non-trivial SCCs and three preferred labellings: ($\{a_1, a_2, a_3, a_4, a_5, a_7, a_{15}\}$, $\{a_8, a_9, a_{10}, a_{12}, a_{13}\}$, $\{a_6, a_{11}, a_{14}\}$), ($\{a_1, a_2, a_3, a_6, a_8, a_{12}, a_{15}\}$, $\{a_4, a_5, a_7, a_9, a_{10}, a_{11}, a_{13}\}$, $\{a_{14}\}$) and ($\{a_2, a_3, a_6, a_8, a_9, a_{12}, a_{13}\}$, $\{a_1, a_4, a_5, a_7, a_{10}, a_{11}, a_{14}, a_{15}\}$, $\{\}$). The execution time of the MC algorithm for this argumentation framework is 183.031 seconds. This example shows again that the number of UNDEC-labelled arguments and the number of labellings are closely related to the time complexity of the MC algorithm. This problem could be effectively coped with by our decomposition-based algorithm.

According to the above results, we may come to the following conclusions:

- When the ratio of the number of edges to the number of nodes of an argumentation framework is about 1:1, the execution time of our algorithm is very low and stable.
- When the ratio of the number of edges to the number of nodes of an argumentation framework is no more than 1.5:1, the execution time of our algorithm is obviously better than the MC algorithm.
- When the ratio of the number of edges to the number of nodes of an argumentation framework is bigger than 4:1, the difference between our algorithm and the existing one is not apparent.
- Two factors that affect the time complexity of the MC algorithm for computing preferred labellings are the number of UNDEC-labelled arguments in a labelling and the number of preferred labellings. Our decomposition-based approach could effectively cope with this problem.

6.5 Conclusions

In this chapter, we have introduced an approach for incrementally computing the semantics of a static argumentation framework, and then performed an empirical investigation.

The decomposition-based approach is realized by exploiting the notion of strongly connected components of a directed graph. Given an argumentation framework (which is regarded as a directed graph), it is firstly decomposed into a set of SCCs, and then into a set of sub-frameworks located in a number of layers. According to the decomposition, under some typical semantics (admissible, complete, grounded, preferred), the semantics of the argumentation framework are computed incrementally from the lowest layer in which each sub-framework is not restricted by other sub-frameworks, to the highest layer in which each sub-framework is most restricted by the sub-frameworks located in the lower layers. The empirical results showed that when the defeat graphs of argumentation frameworks are sparse (more specifically, when the ratio of the number of edges to the number of nodes of a defeat graph is less than 1.5:1), the efficiency of the decomposition-based approach is apparent.

It should be noticed that the efficiency of the decomposition-based approach is highly limited by the maximal SCC of an argumentation framework. In [17], we present a further solution by exploiting the most skeptically rejected arguments of an argumentation framework. Given an argumentation framework, its grounded labelling is first generated. Then, the attacks between the undecided arguments and the rejected arguments are removed. It turns out that the modified argumentation framework has the same preferred labellings as the original argumentation framework, but the maximal SCC in it could be much smaller than that of the original argumentation framework. Empirical results show that this new method dramatically reduces the computation time for some sparse argumentation frameworks (when the ratio of the number of edges to the number of nodes of an AF is between 1:1 and 1.8:1).

Besides the decomposition-based approach presented in this chapter, some other work for efficiently computing the semantics of a static argumentation framework mainly consist of the following three lines.

The first line of work is on identifying tractable classes of argumentation frameworks with special structures [1], and developing efficient algorithms for some classes of argumentation frameworks with fixed parameters, such as bounded tree-width [4] and bounded clique-width [18], etc. And, in [19], Dvořák et al proposed a generic approach for solving hard problems in the area of argumentation in a "complexity-sensitive" way. The corresponding empirical results showed that their approach significantly outperforms existing systems developed for hard argumentation problems (i.e., problems under the preferred, semi-stable or stage semantics).

The second line of work is on decomposition-based computation of extensions [15,20]. In [15], Baroni et al proposed a SCC-recursive scheme for argumentation semantics, based on decomposition along the strongly connected components of an argumentation framework. Meanwhile, they pointed out, at the end of the paper, that "it is worth investigating the development of efficient and incremental algorithms based on local computation at the level of strongly connected components". In line with Baroni et al's proposal, Baumann et al developed splitting-based algorithms for the computation of extensions. Their experimental results showed an average improvement by 50% and by 54% for preferred and stable semantics respectively, compared to Modgil and Caminada's algorithms [6].

The third line of work is to develop specific algorithms to improve the efficiency of computing the status of arguments. For instance, in [21], Nofal et al presented a case study on experimental algorithms in the context of an instance of extended argumentation frameworks, and analysed the efficiency of three different algorithms for deciding the acceptability of an argument with respect to a set of arguments; in [22], they proposed a more efficient algorithm for enumerating all preferred extensions, by utilizing further labels[8] to improve labels' transitions.

The approach proposed in this chapter is related to the above three lines of work, but differs from them in the following ways. Firstly, while the first and third lines of work treat argumentation frameworks as monolithic entities and therefore have not considered the local

---

[8] Besides applying the labels IN, OUT and UNDEC, they introduced the use of MUST-OUT and IGNORED, in which the MUST-OUT label identifies arguments that attack IN arguments. The IGNORED label designates arguments which might not be included in a preferred extension because they might not be defended by any IN argument.

tractability of a generic argumentation framework, our decomposition-based approach takes advantage of the local tractability of acyclic fragments. Secondly, the similarity between our work and the second line of work is the idea of incremental computation of argumentation semantics. However, the existing works have not considered how to exploit the tractability of some sub-frameworks with special topologies (in this book acyclic sub-frameworks). Thirdly, the layered approach for decomposing an argumentation framework and combining argumentation semantics has not been introduced in the existing literature.

References

[1] P. Dunne, Computational properties of argument systems satisfying graph-theoretic constraints, Artificial Intelligence 171 (10–15) (2007) 701–729.

[2] P. Dung, R. Kowalski, F. Toni, Dialectic proof procedures for assumption-based, admissible argumentation, Artificial Intelligence 170 (2) (2006) 114–159.

[3] S. Coste-Marquis, C. Devred, P. Marquis, Symmetric argumentation frameworks, in: Proceedings of the Eighth European Conferences on Symbolic and Quantitative Approaches to Reasoning with Uncertainty, 2005, pp. 317–328.

[4] W. Dvořák, R. Pichler, S. Woltran, Towards fixed-parameter tractable algorithms for argumentation, in: Proceedings of the 12th International Conference on the Principles of Knowledge Representation and Reasoning, 2010, pp. 112–122.

[5] C. Cayrol, S. Doutre, J. Mengin, On decision problems related to the preferred semantics for argumentation frameworks, Journal of Logic and Computation 13 (3) (2003) 377–403.

[6] S. Modgil, M. Caminada, Proof theories and algorithms for abstract argumentation frameworks, Argumentation in Artificial Intelligence (2009) 105–129.

[7] P. Dung, P. Thang, A sound and complete dialectical proof procedure for sceptical preferred argumentation, in: Proceedings of an LPNMR Workshop on Argumentation and Non-Monotonic Reasoning, 2007, pp. 49–63.

[8] P. Dung, P. Mancarella, F. Toni, Computing ideal sceptical argumentation, Artificial Intelligence 171 (10–15) (2007) 642–674.

[9] U. Egly, S. Gaggl, S. Woltran, Aspartix: Implementing argumentation frameworks using answer-set programming, in: Proceedings of the 24th International Conference on Logic Programming, 2009, pp. 734–738.

[10] J. Nieves, U. Cortés, M. Osorio, Preferred extensions as stable models, Theory and Practice of Logic Programming 8 (4) (2008) 527–543.

[11] T. Wakaki, K. Nitta, Computing argumentation semantics in answer set programming, in: Proceedings of the 22nd Annual Conference of the Japanese Society for Artificial Intelligence, 2008, pp. 254–269.

[12] E. Kim, S. Ordyniak, S. Szeider, Algorithms and complexity results for persuasive argumentation, Artificial Intelligence 175 (9–10) (2011) 1722–1736.

[13] S. Coste-Marquis, C. Devred, M. Lagasquie-Schiex, P. Marquis, On the merging of dung's argumentation systems, Artificial Intelligence 171 (10–15) (2007) 730–753.

[14] R. Tarjan, Depth-first search and linear graph algorithms, SIAM Journal on Computing 1 (2) (1972) 146–160.

[15] P. Baroni, M. Giacomin, G. Guida, Scc-recursiveness: a general schema for argumentation semantics, Artificial Intelligence 168 (1–2) (2005) 162–210.

[16] S. Dasgupta, C.H. Papadimitriou, U.V. Vazirani, Algorithms, McGraw-Hill, 2006.

[17] B. Liao, L. Lei, J. Dai, Computing preferred labellings by exploiting SCCs and most sceptically rejected arguments, in: Proceedings of Second International Workshop on Theory and Applications of Formal Argumentation, 2013.

[18] W. Dvořák, S. Szeider, S. Woltran, Reasoning in argumentation frameworks of bounded clique-width, in: Proceedings of the Third International Conference on Computational Models of Argument, 2010, pp. 219–230.

[19] W. Dvořák, M. Järvisalo, J.P. Wallner, S. Woltran, Complexity-sensitive decision procedures for abstract argumentation, in: Proceedings of the 13th International Conference on the Principles of Knowledge Representation and Reasoning, 2012, pp. 54–64.

[20] R. Baumann, G. Brewka, R. Wong, Splitting argumentation frameworks: an empirical evaluation, in: Proceedings of the First International Workshop on Theory and Applications of Formal Argumentation, 2011, pp. 17–31.

[21] S. Nofal, P. Dunne, K. Atkinson, Towards experimental algorithms for abstract argumentation, in: Proceedings of the Fourth International Conference on Computational Models of Argument, 2012, pp. 217–228.

[22] S. Nofal, P. Dunne, K. Atkinson, On preferred extension enumeration in abstract argumentation, in: Proceedings of the Fourth International Conference on Computational Models of Argument, 2012, pp. 205–216.

CHAPTER 7

An Approach for Dynamic Argumentation Frameworks

Chapter Outline

7.1 Introduction

According to the existing literature, most argumentation systems are dynamic [1–3], especially argumentation-based autonomous agents within a dynamic environment, including belief revision [4–9], deliberation [10–13], decision-making [14–18], and negotiation [19–23]. The existing research shows that in many argumentation systems, arguments and their attack relation evolve with the changing of underlying knowledge or information. For example, in [6,7], the authors formulated a system where an instantiated argumentation framework is based on the changing observations. So, at each time point, when observations change, the arguments and their attack relation change accordingly. In [4,24], within an argumentative system, when a new explanation is received, some strict rules are changed to defeasible rules, which gives rise to the changing of arguments and their attack relation. In [13,25], due to the dynamics of observations and inference rules, the argumentation frameworks for beliefs, goals and intentions, respectively, are dynamic. In [22,23], argumentation-based negotiation (ABN) agents perform reasoning with incomplete, uncertain and inconsistent information. Each agent's theory (as an argumentation system) may evolve during a negotiation dialogue, i.e., if an agent receives an argument from another agent, it will add the new argument to its theory, and furthermore, new conflicts may arise between the original arguments of the agent and the ones that emerge after adding the received arguments to its theory [23]. In [26], when a collection of argumentation systems coming from different agents are merged (after

Efficient Computation of Argumentation Semantics. http://dx.doi.org/10.1016/B978-0-12-410406-8.00007-5

consensual expansion of each argumentation system), the arguments and the attack relation of each argumentation system may change accordingly.

To illustrate the dynamics of argumentation systems, let us see a revised example from [6].

Example 7.1. There are some rules in the knowledge base of an autonomous agent for performing basic email filtering, in which r_4 is superior to r_5, and r_7 is superior to r_5:

r_1: A message from the local host is usually not classified as spam.
r_2: A message is usually labelled as spam if it comes from a server that is on the blacklist.
r_3: A message should be moved to the "junk" folder if it is marked as spam.
r_4: Unfiltered messages in the "junk" folder usually should not be moved to the inbox.
r_5: If an email does not match with any user-defined filter then it usually should be moved to the "inbox" folder.
r_6: A message should not be moved to the "junk" folder if it is from a VIP user.
r_7: A message with viruses should not be moved to the inbox.

Based on the above rules, let us consider the following scenarios at successive time points t_1, t_2, t_3 and t_4. First, at t_1, the observations related to a message M_1 show that: (1) M_1 is from a local host, and (2) M_1 comes from a server that is in the blacklist. Second, at t_2, a new observation "M_1 does not match with any user-defined filter" is added. Third, at t_3, an observation "M_1 is from a VIP user" is added. Fourth, at t_4, an observation "M_1 contains a virus" is added. So, according to the rules $r_1 \sim r_7$, at different time points, we may construct the following four different argumentation frameworks (F_1, F_2, F_3, and F_4), in which (α, β) denotes α rebuts β,[1] while $\langle \alpha, \beta \rangle$ denotes α undercuts β (i.e., $\exists \gamma \subset \beta$, such that, γ rebuts α) [25].

$$F_1 = (\{a_1, a_2, a_3, a_4\}, \{\langle a_1, a_3 \rangle, (a_1, a_2), (a_2, a_1), \langle a_1, a_4 \rangle\})$$

$$F_2 = (\{a_1, a_2, a_3, a_4, a_5\}, \{\langle a_1, a_3 \rangle, (a_1, a_2), (a_2, a_1), \langle a_1, a_4 \rangle, (a_4, a_5)\})$$

$$F_3 = (\{a_1, a_2, a_3, a_4, a_5, a_6\}, \{\langle a_1, a_3 \rangle, (a_1, a_2), (a_2, a_1), \langle a_1, a_4 \rangle, (a_4, a_5),$$
$$(a_3, a_6), (a_6, a_3), \langle a_6, a_4 \rangle\})$$

$$F_4 = (\{a_1, a_2, a_3, a_4, a_5, a_6, a_7\}, \{\langle a_1, a_3 \rangle, (a_1, a_2), (a_2, a_1),$$
$$\langle a_1, a_4 \rangle, (a_4, a_5), (a_3, a_6), (a_6, a_3), \langle a_6, a_4 \rangle, (a_7, a_5)\})$$

$$a_1 = (\{r_1\}, \neg spam(M_1))$$

$$a_2 = (\{r_2\}, spam(M_1))$$

$$a_3 = (\{r_2, r_3\}, move\_junk(M_1))$$

$$a_4 = (\{r_2, r_3, r_4\}, \neg move\_inbox(M_1))$$

$$a_5 = (\{r_5\}, move\_inbox(M_1))$$

---

[1] When the conclusions of arguments α and β are complementary, if α is superior to β, then α is a proper defeater of β, denoted as (α, β); else, α is a blocking defeater of β, denoted as $\{(\alpha, \beta), (\beta, \alpha)\}$. For example, since r_4 is superior to r_5, the corresponding argument a_4 is superior to a_5. So, there is only an attack from a_4 to a_5, and not vice versa.

$$a_6 = (\{r_6\}, \neg move\_junk(M_1))$$
$$a_7 = (\{r_7\}, \neg move\_inbox(M_1))$$

In Figure 7.1, the status of arguments a_1, a_2, a_3, and a_4 in F_1 remains unchanged after the addition of arguments a_5, and of attacks (a_4, a_5), in F_2; the status of arguments a_1 and a_2 in F_2 remains unchanged, while the status of arguments a_3, a_4, and a_5 in F_2 may be affected, after the addition of an argument a_6, and of attacks (a_3, a_6), (a_6, a_3), and $\langle a_6, a_4 \rangle$, in F_3; the status of arguments a_1, a_2, a_3, a_4, and a_6 in F_3 remains unchanged, while the status of arguments a_5 may be affected, after the addition of argument a_7, and of the attack (a_7, a_5).

As shown in Example 7.1, when new observations arise, a set of arguments and/or attacks are *added* to the system. Meanwhile, in some other cases, when observations or rules of inference change, a set of arguments and/or attacks may be *deleted* from the system [25]. With the changing of arguments and/or attacks of an argumentation system, the status of some arguments changes, while that of others remains untouched. Now, one of the challenging problems is how to efficiently compute the dynamics of argumentation systems. When an argumentation system is modified, we may simply recompute the status of each argument afresh. However, this method is obviously inefficient, and in most of cases, difficult.

To cope with this problem, there have been a small number of efforts [16,27,28]. First, Boella et al studied the dynamics of argumentation by exploring the principles according to which the extension does not change when the set of arguments or the attacks between them are changed [27]. However, they have not considered how the extensions of an argumentation system evolve when new arguments are added or the old ones are removed. Second, Cayrol et al addressed the problem of revising the set of extensions of an abstract argumentation system, and studied how the extensions of an argumentation system may evolve when a new argument is received [28]. However, they restricted their study to the case of adding just *one* argument having only *one* interaction with an initial argument. Third, Amgoud et al used dynamics of argumentation in the decision-making of an autonomous agent [16]. They studied how the

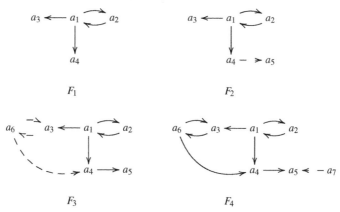

Figure 7.1 The evolution on an argumentation system.

acceptability of arguments evolves when the decision system is extended by new arguments without computing the whole extensions. However, they also considered the situation where only *one* argument is added to the system. In addition, all of the current efforts have not taken the efficiency of computing the dynamics of argumentation into consideration.

According to the above analysis, a more general theory is needed to formulate the dynamics of an argumentation system, with the following three characteristics:

- The number of arguments and attacks to be added to (or deleted from) an argumentation system is unlimited;
- Both the addition and the deletion of arguments and/or attacks are considered;
- The efficiency of computing the dynamics of argumentation systems is considered.

To realize this purpose, we introduce a division-based approach. The basic idea of this method is that when an argumentation framework is updated, we may only recompute the status of those arguments that are *affected*, while the status of those arguments that are *unaffected* can be obtained directly from the result of the previous computation. Based on the fundamental theories presented in Part III, the status of arguments that are *affected* and *unaffected* can be computed separately and then combined to form the semantics of the updated argumentation framework.

7.2   The Changing of an Argumentation Framework

As mentioned in the previous section, when a set of arguments and/or a set of attacks are added to (or deleted from) an argumentation framework, the status of affected arguments may change. However, if some arguments and attacks to be added are in the existing argumentation framework, or some arguments and attacks to be deleted are not in the existing argumentation framework, an addition (respectively a deletion) of them will not have any effects on the argumentation framework. So, for simplicity and without loss of generality, we suppose that the arguments and attacks to be added are not in the existing argumentation framework, while the arguments and attacks to be deleted are in the existing argumentation framework.

Let U_{arg} be the universe of arguments. For all $A \subseteq U_{arg}$ and $B \subseteq U_{arg} \setminus A$, let $\Im_{A:B}$ be the set of attacks (also called interactions) related to B and of the form (α, β), (β, α), or (β, β'), in which $\alpha \in A$ and $\beta, \beta' \in B$. Let $\Im_A$ be a set of attacks between the arguments in A, and of the form (α, α'), in which $\alpha, \alpha' \in A$. An addition of an argumentation framework is defined as follows.

Definition 7.1 (An addition of an argumentation framework).   Let $F = (A, R)$ be an argumentation framework, where $A \subseteq U_{arg}$ and $R \subseteq A \times A$. An addition of F is represented as a tuple $(B, \Im_{A:B} \cup \Im_A)$, in which $B \subseteq U_{arg} \setminus A$ is a set of arguments to be added, and $\Im_{A:B} \cup \Im_A$ is a set of attacks to be added.

Meanwhile, let $\mathfrak{I}_{A\backslash B:B}$ and $\mathfrak{I}_{A\backslash B}$ be sets of attacks to be deleted. It holds that $\mathfrak{I}_{A\backslash B:B} = R \cap ((B \times B) \cup (B \times (A \backslash B)) \cup ((A \backslash B) \times B)) = R \cap ((B \times A) \cup (A \times B))$. A deletion of an argumentation framework is defined as follows.

Definition 7.2 (A deletion of an argumentation framework). Let $F = (A, R)$ be an argumentation framework, where $A \subseteq \mathsf{U}_{arg}$ and $R \subseteq A \times A$. A deletion of F is also represented as a tuple $(B, \mathfrak{I}_{A\backslash B:B} \cup \mathfrak{I}_{A\backslash B})$, in which $B \subseteq A$ is a set of arguments to be deleted, and $\mathfrak{I}_{A\backslash B:B} \cup \mathfrak{I}_{A\backslash B}$ is a set of attacks to be deleted.

Based on Definitions 7.1 and 7.2, an updated argumentation framework with respect to an addition or a deletion is syntactically defined as follows:

Definition 7.3 (Updated argumentation framework). Let $F = (A, R)$ be an argumentation framework, $(B, \mathfrak{I}_{A:B} \cup \mathfrak{I}_A)$ be an addition and $(B, \mathfrak{I}_{A\backslash B:B} \cup \mathfrak{I}_{A\backslash B})$ be a deletion. Syntactically, the updated argumentation framework with respect to $(B, \mathfrak{I}_{A:B} \cup \mathfrak{I}_A)$ and $(B, \mathfrak{I}_{A\backslash B:B} \cup \mathfrak{I}_{A\backslash B})$ is respectively represented as follows:

$$(A, R) \oplus (B, \mathfrak{I}_{A:B} \cup \mathfrak{I}_A) =_{def} (A \cup B, R \cup (\mathfrak{I}_{A:B} \cup \mathfrak{I}_A)) \tag{7.1}$$

$$(A, R) \ominus (B, \mathfrak{I}_{A\backslash B:B} \cup \mathfrak{I}_{A\backslash B}) =_{def} (A \backslash B, R \backslash (\mathfrak{I}_{A\backslash B:B} \cup \mathfrak{I}_{A\backslash B})) \tag{7.2}$$

For simplicity, we use $(A^{\oplus}, R^{\oplus})$ and $(A^{\ominus}, R^{\ominus})$ to denote $(A \cup B, R \cup (\mathfrak{I}_{A:B} \cup \mathfrak{I}_A))$ and $(A \backslash B, R \backslash (\mathfrak{I}_{A\backslash B:B} \cup \mathfrak{I}_{A\backslash B}))$, respectively.

Proposition 7.1. *Let $(A^{\oplus}, R^{\oplus})$ and $(A^{\ominus}, R^{\ominus})$ be the result of an addition and of a deletion respectively. It holds $R^{\oplus} \subseteq A^{\oplus} \times A^{\oplus}$ and $R^{\ominus} \subseteq A^{\ominus} \times A^{\ominus}$, i.e. $(A^{\oplus}, R^{\oplus})$ and $(A^{\ominus}, R^{\ominus})$ are argumentation frameworks.*

Proof. First, according to Definitions 7.1 and 7.2, in the case of addition, it holds that $R \subseteq A \times A, \mathfrak{I}_{A:B} \subseteq (A \times B) \cup (B \times A) \cup (B \times B)$, and $\mathfrak{I}_A \subseteq A \times A$. So, we have $R^{\oplus} = R \cup (\mathfrak{I}_{A:B} \cup \mathfrak{I}_A) \subseteq (A \times A) \cup ((A \times B) \cup (B \times A) \cup (B \times B)) \cup (A \times A) = (A \cup B) \times (A \cup B) = A^{\oplus} \times A^{\oplus}$. Second, in the case of deletion, it holds that $\mathfrak{I}_{A\backslash B:B} = R \cap ((B \times B) \cup (B \times (A \backslash B)) \cup ((A \backslash B) \times B))$. Therefore, $R \backslash \mathfrak{I}_{A\backslash B:B} = (R \cap (A \times A)) \backslash (R \cap ((B \times B) \cup (B \times (A \backslash B)) \cup ((A \backslash B) \times B))) = R \cap ((A \backslash B) \times (A \backslash B))$. Meanwhile, since all interactions in $\mathfrak{I}_{A\backslash B}$ belong to $R \cap ((A \backslash B) \times (A \backslash B))$, it holds that $(R \backslash \mathfrak{I}_{A\backslash B:B}) \backslash \mathfrak{I}_{A\backslash B} \subseteq (A \backslash B) \times (A \backslash B) = A^{\ominus} \times A^{\ominus}$, i.e., $R^{\ominus} = R \backslash (\mathfrak{I}_{A\backslash B:B} \cup \mathfrak{I}_{A\backslash B}) = (R \backslash \mathfrak{I}_{A\backslash B:B}) \backslash \mathfrak{I}_{A\backslash B} \subseteq A^{\ominus} \times A^{\ominus}$. $\square$

7.3 The Division of an Updated Argumentation Framework

According to Formulas 7.1 and 7.2, with respect to a certain argumentation semantics σ, we may directly compute the extensions of $(A^{\oplus}, R^{\oplus})$ and $(A^{\ominus}, R^{\ominus})$ afresh, without considering previous information (the extensions of $F = (A, R)$). However, if the previous information is properly used, the complexity of computing the dynamics of argumentation might be decreased. If an argumentation semantics satisfies the criterion of *directionality* [29], then the status of an argument α is affected only by the status of its defeaters, while the arguments which only receive an attack from α have no effect on the state of α. Based on this idea, we may infer that when a set of arguments and/or a set of attacks are added to (or deleted from) an argumentation framework, some arguments are directly or indirectly affected, while others remain unaffected. Therefore, if we can divide $(A^{\oplus}, R^{\oplus})$ or $(A^{\ominus}, R^{\ominus})$ into two parts: *affected* and *unaffected*, we need only to compute the status of affected arguments, while leaving the status of unaffected arguments unchanged. According to the theories presented in Chapters 4 and 5, the status of affected arguments can be computed in a conditioned sub-framework, and then combined with that of the unaffected ones, to form the semantics of the updated argumentation framework. Now, let us introduce the notion of dividing an updated argumentation framework.

The division of an updated argumentation framework is based on the notion of *directionality* of argumentation semantics [29,30]. Given an argumentation framework $F = (A, R)$, for all $\alpha, \beta \in A$, if there exists a directed path from α to β, i.e, β is reachable from α, then under the semantics that satisfies directionality, the status of β may be affected by α; otherwise, β is independent of α. Based on this idea, the notion of *reachability*, as well as the notions of *affected* and *unaffected* between two arguments can be defined as follows:

Definition 7.4 (Reachability of two arguments).   Let $\alpha, \beta \in U_{arg}$ be two arguments, and $R \subseteq U_{arg} \times U_{arg}$ be a set of attacks. The *reachability* of two arguments with respect to R is recursively defined as follows:

- If $(\alpha, \beta) \in R$, then β is reachable from α;
- If $\exists \gamma \in U_{arg}$, such that γ is reachable from α with respect to R and $(\gamma, \beta) \in R$, then β is reachable from α.

According to Definition 7.4, the *reachability* relation is transitive, i.e., if γ is reachable from β and β is reachable from α, then γ is reachable from α.

Definition 7.5 (Affected and unaffected between two arguments).   Let $\alpha, \beta \in U_{arg}$ be two arguments, and $R \subseteq U_{arg} \times U_{arg}$ be a set of attacks. We say that under the semantics that satisfies directionality, the status of β is affected by α, if and only if β is reachable from α with respect to R. Otherwise, β is unaffected by α with respect to R.

Based on Definition 7.5, given a set of arguments $A \subseteq U_{arg}$, it is possible to identify the subset of A that is affected by a set of arguments $B \subseteq U_{arg}$ or by a set of attacks (interactions) $I \subseteq U_{arg} \times U_{arg}$, with respect to a set of attacks $R \subseteq U_{arg} \times U_{arg}$. In addition, we notice that the set of affected arguments related to an attack $r = (\alpha, \beta)$ can be computed through β, the attacked argument of r. So, we will define the affected and unaffected arguments related to an attack r by using the attacked argument of r. Here, we use $attacked(r)$ to indicate the attacked argument of r, and $attacked(I) = \{attacked(r) \mid r \in I\}$ to indicate a set of arguments, each of which is the attacked argument of an attack in I.

Definition 7.6 (Affected arguments). Let $A, B \subseteq U_{arg}$ be sets of arguments, and $R, I \subseteq U_{arg} \times U_{arg}$ be sets of attacks. We use $afct(A, B, R)$ and $afct(A, I, R)$ to indicate the set of arguments within A that are affected respectively by B and I, with respect to R. Formally, we have:

$$afct(A, B, R) = \bigcup_{\alpha \in B} \{\beta \in A : \beta \text{ is reachable from } \alpha \text{ with respect to } R\} \qquad (7.3)$$

$$afct(A, I, R) = attacked(I) \cup afct(A, attacked(I), R) \qquad (7.4)$$

Based on the concept of *affected* and *unaffected* arguments, we are ready to define the concept of *the division of an updated argumentation framework*. When an addition $(B, \Im_{A:B} \cup \Im_A)$ is added to (or a deletion $(B, \Im_{A\setminus B:B} \cup \Im_{A\setminus B})$ is deleted from) an argumentation framework $F = (A, R)$, F will be divided into three parts:

- a component of F that is affected by $(B, \Im_{A:B} \cup \Im_A)$ (respectively $(B, \Im_{A\setminus B:B} \cup \Im_{A\setminus B})$),
- a component of F that is unaffected by $(B, \Im_{A:B} \cup \Im_A)$ (respectively $(B, \Im_{A\setminus B:B} \cup \Im_{A\setminus B})$), and
- a subset of the unaffected component that conditions the affected arguments.

Formally, we first present the definition of the division of an updated argumentation framework with respect to an addition $(B, \Im_{A:B} \cup \Im_A)$.

Definition 7.7 (The division of an updated framework with respect to an addition). Let $F = (A, R)$ be an argumentation framework, and $(B, \Im_{A:B} \cup \Im_A)$ be an addition to it. The updated argumentation framework $(A^{\oplus}, R^{\oplus})$ is divided into three parts: $(A_a^{\oplus}, R_a^{\oplus})$, $(A_u^{\oplus}, R_u^{\oplus})$, and $(A_c^{\oplus}, R_c^{\oplus})$, where a, u, and c stand for, respectively, *affected*, *unaffected* and *conditioning*.

$$A_a^{\oplus} = afct(A, B, R \cup \Im_{A:B}) \cup afct(A, \Im_A, R) \cup B$$
$$A_u^{\oplus} = A \setminus (afct(A, B, R \cup \Im_{A:B}) \cup afct(A, \Im_A, R))$$
$$A_c^{\oplus} = \{\beta \in A_u^{\oplus} \mid \exists \alpha \in A_a^{\oplus}, \text{ such that } (\beta, \alpha) \in R \cup \Im_{A:B} \cup \Im_A\}$$
$$R_a^{\oplus} = (R \cup \Im_A \cup \Im_{A:B}) \cap (A_a^{\oplus} \times A_a^{\oplus})$$
$$R_u^{\oplus} = R \cap (A_u^{\oplus} \times A_u^{\oplus})$$
$$R_c^{\oplus} = (R \cup \Im_A \cup \Im_{A:B}) \cap (A_c^{\oplus} \times A_a^{\oplus})$$

In this definition, the set of affected arguments $A_a^{\oplus}$ contains those arguments in A that are affected by B and $\Im_A$, as well as those arguments in B. $A_u^{\oplus}$ is the set of arguments in $A^{\oplus}$ that are unaffected. The set of conditioning arguments $A_c^{\oplus}$ contains those arguments in $A_u^{\oplus}$ that attack the arguments in $A_a^{\oplus}$.

In order to ensure the correctness of the division defined in Definition 7.7, in this stage, we should verify that: (1) the union of affected arguments and the unaffected ones is equal to the set of arguments in the updated argumentation framework, i.e., $A_a^\oplus \cup A_u^\oplus = A^\oplus$ (obvious), and (2) the union of attacks in three parts (affected, unaffected, conditioning) is equal to the set of attacks in the updated argumentation framework, which is formulated by the following lemma:

Lemma 7.1.   *It holds that* $R_c^\oplus \cup R_a^\oplus \cup R_u^\oplus = R \cup \mathfrak{I}_A \cup \mathfrak{I}_{A:B}$.

Proof.   Firstly, we identify the characteristics of the relations among A, $A_u^\oplus$, $A_c^\oplus$, and $A_a^\oplus$: since arguments in $A_a^\oplus$ do not attack arguments in $A_u^\oplus$, it holds that $(R \cup \mathfrak{I}_A \cup \mathfrak{I}_{A:B}) \cap (A_a^\oplus \times A_u^\oplus) = \emptyset$; since in $A_u^\oplus$, only arguments in $A_c^\oplus \subseteq A_u^\oplus$ attack $A_a^\oplus$, it holds that $(R \cup \mathfrak{I}_A \cup \mathfrak{I}_{A:B}) \cap (A_u^\oplus \times A_a^\oplus) = (R \cup \mathfrak{I}_A \cup \mathfrak{I}_{A:B}) \cap (A_c^\oplus \times A_a^\oplus)$; according to $A_u^\oplus = A \setminus (afct(A, B, R \cup \mathfrak{I}_{A:B}) \cup afct(A, \mathfrak{I}_A, R))$ and $A_a^\oplus = afct(A, B, R \cup \mathfrak{I}_{A:B}) \cup afct(A, \mathfrak{I}_A, R) \cup B$, it holds that $A_u^\oplus \cup A_a^\oplus = A \cup B$. In addition, since $(\mathfrak{I}_A \cup \mathfrak{I}_{A:B}) \cap (A_u^\oplus \times A_u^\oplus) = \emptyset$, it holds that $(R \cup \mathfrak{I}_A \cup \mathfrak{I}_{A:B}) \cap (A_u^\oplus \times A_u^\oplus) = R \cap (A_u^\oplus \times A_u^\oplus)$. Secondly, according to these relations, it holds that:

$$
\begin{aligned}
R_c^\oplus \cup R_a^\oplus \cup R_u^\oplus &= ((R \cup \mathfrak{I}_A \cup \mathfrak{I}_{A:B}) \cap (A_c^\oplus \times A_a^\oplus)) \cup \\
&\quad ((R \cup \mathfrak{I}_A \cup \mathfrak{I}_{A:B}) \cap (A_a^\oplus \times A_a^\oplus)) \cup (R \cap (A_u^\oplus \times A_u^\oplus)) \\
&= (R \cup \mathfrak{I}_A \cup \mathfrak{I}_{A:B}) \cap ((A_a^\oplus \times A_a^\oplus) \cup (A_u^\oplus \times A_u^\oplus) \cup (A_c^\oplus \times A_a^\oplus)) \\
&= (R \cup \mathfrak{I}_A \cup \mathfrak{I}_{A:B}) \\
&\quad \cap ((A_a^\oplus \times A_a^\oplus) \cup (A_u^\oplus \times A_u^\oplus) \cup (A_u^\oplus \times A_a^\oplus) \cup (A_a^\oplus \times A_u^\oplus)) \\
&= (R \cup \mathfrak{I}_A \cup \mathfrak{I}_{A:B}) \cap ((A_u^\oplus \cup A_a^\oplus) \times (A_u^\oplus \cup A_a^\oplus)) \\
&= (R \cup \mathfrak{I}_A \cup \mathfrak{I}_{A:B}) \cap ((A \cup B) \times (A \cup B)) \\
&= R \cup \mathfrak{I}_A \cup \mathfrak{I}_{A:B}. \qquad \square
\end{aligned}
$$

Similar to Definition 7.7, for the division of an argumentation framework with respect to a deletion, we have the following definition and lemma:

Definition 7.8 (The division of an updated framework with respect to a deletion).   Let $F = (A, R)$ be an argumentation framework, and $(B, \mathfrak{I}_{A\setminus B:B} \cup \mathfrak{I}_{A\setminus B})$ be a deletion to it. The updated argumentation framework $(A^\ominus, R^\ominus)$ is divided into three parts: $(A_a^\ominus, R_a^\ominus)$, $(A_u^\ominus, R_u^\ominus)$, and $(A_c^\ominus, R_c^\ominus)$.

$$
\begin{aligned}
A_a^\ominus &= afct(A \setminus B, B, R) \cup afct(A \setminus B, \mathfrak{I}_{A\setminus B}, R) \\
A_u^\ominus &= A \setminus (A_a^\ominus \cup B) \\
A_c^\ominus &= \{\beta \in A_u^\ominus \mid \exists \alpha \in A_a^\ominus, \text{ such that } (\beta, \alpha) \in R \setminus \mathfrak{I}_{A\setminus B}\} \\
R_a^\ominus &= (R \setminus \mathfrak{I}_{A\setminus B}) \cap (A_a^\ominus \times A_a^\ominus) \\
R_u^\ominus &= R \cap (A_u^\ominus \times A_u^\ominus) \\
R_c^\ominus &= (R \setminus \mathfrak{I}_{A\setminus B}) \cap (A_c^\ominus \times A_a^\ominus)
\end{aligned}
$$

In this definition, the set of affected arguments in $A_a^\ominus$ are those arguments in $A \setminus B$ (i.e., $A^\ominus$), that are affected by B and $\mathfrak{I}_{A \setminus B}$. $A_u^\ominus$ is the set of arguments that are unaffected. The set of conditioning arguments $A_c^\ominus$ contains those arguments in $A_u^\ominus$ that attack the arguments in $A_a^\ominus$.

Lemma 7.2. *It holds that* $R_c^\ominus \cup R_a^\ominus \cup R_u^\ominus = R \setminus (\mathfrak{I}_{A \setminus B:B} \cup \mathfrak{I}_{A \setminus B})$.

Proof. Similar to Lemma 7.1, we have: since $\mathfrak{I}_{A \setminus B} \cap (A_u^\ominus \times A_u^\ominus) = \emptyset$, it holds that $R \cap (A_u^\ominus \times A_u^\ominus) = (R \setminus \mathfrak{I}_{A \setminus B}) \cap (A_u^\ominus \times A_u^\ominus)$; since in $A_a^\ominus$, only arguments in $A_c^\ominus \subseteq A_u^\ominus$ attack $A_a^\ominus$, it holds that $(R \setminus \mathfrak{I}_{A \setminus B}) \cap (A_u^\ominus \times A_a^\ominus) = (R \setminus \mathfrak{I}_{A \setminus B}) \cap (A_c^\ominus \times A_a^\ominus)$; since arguments in $A_a^\ominus$ do not attack arguments in $A_u^\ominus$, it holds that $(R \setminus \mathfrak{I}_{A \setminus B}) \cap (A_a^\ominus \times A_u^\ominus) = \emptyset$; since interactions in $\mathfrak{I}_{A \setminus B}$ are not related to B, it holds that $\mathfrak{I}_{A \setminus B} \cap ((A \times B) \cup (B \times A)) = \emptyset$; according to $A_u^\ominus = A \setminus (A_a^\ominus \cup B)$, it holds that $A_a^\ominus \cup A_u^\ominus = A \setminus B$; and according to Definition 7.2, $\mathfrak{I}_{A \setminus B:B} = R \cap ((B \times A) \cup (A \times B))$. So, we have:

$$
\begin{aligned}
R_c^\ominus \cup R_a^\ominus \cup R_u^\ominus &= ((R \setminus \mathfrak{I}_{A \setminus B}) \cap (A_c^\ominus \times A_a^\ominus)) \cup \\
&\quad ((R \setminus \mathfrak{I}_{A \setminus B}) \cap (A_a^\ominus \times A_a^\ominus)) \cup (R \cap (A_u^\ominus \times A_u^\ominus)) \\
&= (R \setminus \mathfrak{I}_{A \setminus B}) \cap ((A_a^\ominus \times A_a^\ominus) \cup (A_u^\ominus \times A_u^\ominus) \cup (A_c^\ominus \times A_a^\ominus)) \\
&= (R \setminus \mathfrak{I}_{A \setminus B}) \cap ((A_a^\ominus \times A_a^\ominus) \cup (A_u^\ominus \times A_u^\ominus) \cup (A_u^\ominus \times A_a^\ominus) \cup (A_a^\ominus \times A_u^\ominus)) \\
&= (R \setminus \mathfrak{I}_{A \setminus B}) \cap ((A_a^\ominus \cup A_u^\ominus) \times (A_a^\ominus \cup A_u^\ominus)) \\
&= (R \setminus \mathfrak{I}_{A \setminus B}) \cap ((A \setminus B) \times (A \setminus B)) \\
&= (R \setminus \mathfrak{I}_{A \setminus B}) \cap ((A \times A) \setminus ((A \times B) \cup (B \times A))) \\
&= ((R \setminus \mathfrak{I}_{A \setminus B}) \cap (A \times A)) \setminus ((R \setminus \mathfrak{I}_{A \setminus B}) \cap ((A \times B) \cup (B \times A))) \\
&= (R \setminus \mathfrak{I}_{A \setminus B}) \setminus \mathfrak{I}_{A \setminus B:B} \\
&= R \setminus (\mathfrak{I}_{A \setminus B:B} \cup \mathfrak{I}_{A \setminus B}) \qquad \square
\end{aligned}
$$

After division, we will construct two sub-frameworks of the updated argumentation framework $(A^\oplus, R^\oplus)$ (respectively, $(A^\ominus, R^\ominus)$): an unconditioned sub-framework and a conditioned sub-framework. First, the unconditioned sub-framework of $(A^\oplus, R^\oplus)$ is $(A_u^\oplus, R_u^\oplus)$, while the conditioned sub-framework of $(A^\oplus, R^\oplus)$ with respect to $(A_u^\oplus, R_u^\oplus)$ is constructed according to $(A_a^\oplus, R_a^\oplus)$ and $(A_c^\oplus, R_c^\oplus)$ as follows:

$$CSF_{add} = (A_a^\oplus \cup A_c^\oplus, R_a^\oplus \cup R_c^\oplus) \tag{7.5}$$

According to Definition 7.7, we may infer that: (i) $A_a^\oplus \cap A_u^\oplus = \emptyset$, (ii) $A_c^\oplus \subseteq A_u^\oplus$, and $\forall \alpha \in A_c^\oplus, \exists \beta \in A_a^\oplus$, such that α attacks β, and (iii) $R_c^\oplus \subseteq A_c^\oplus \times A_a^\oplus$. In other words, CSF_{add} is a conditioned sub-framework.

Similarly, we may construct two sub-frameworks of the updated argumentation framework $(A^\ominus, R^\ominus)$: a unconditioned sub-framework $(A_u^\ominus, R_u^\ominus)$, and a conditioned sub-framework with respect to $(A_u^\ominus, R_u^\ominus)$, which is defined as follows:

$$CSF_{del} = (A_a^\ominus \cup A_c^\ominus, R_a^\ominus \cup R_c^\ominus) \tag{7.6}$$

According to Definition 7.8, we may infer that: (i) $A_a^\ominus \cap A_u^\ominus = \emptyset$, (ii) $A_c^\ominus \subseteq A_u^\ominus$, and $\forall \alpha \in A_c^\ominus, \exists \beta \in A_a^\ominus$, such that α attacks β, and (iii) $R_c^\ominus \subseteq A_c^\ominus \times A_a^\ominus$. So, CSF_{del} is also a conditioned sub-framework.

Finally, based on Lemmas 7.1 and 7.2, the following proposition shows the correctness of the division defined in Definitions 7.7 and 7.8, respectively.

Proposition 7.2.   *Syntactically, the result of combining $(A_u^\oplus, R_u^\oplus)$ and $CSF_{add} = (A_a^\oplus \cup A_c^\oplus, R_a^\oplus \cup R_c^\oplus)$ is equal to $(A^\oplus, R^\oplus) = (A \cup B, R \cup (\Im_{A:B} \cup \Im_A))$, i.e., $A_u^\oplus \cup A_a^\oplus = A \cup B$, and $R_u^\oplus \cup R_a^\oplus \cup R_c^\oplus = R \cup \Im_{A:B} \cup \Im_A$; while the result of combining $(A_u^\ominus, R_u^\ominus)$ and $CSF_{del} = (A_a^\ominus \cup A_c^\ominus, R_a^\ominus \cup R_c^\ominus)$ is equal to $(A^\ominus, R^\ominus) = (A \setminus B, R \setminus (\Im_{A \setminus B:B} \cup \Im_{A \setminus B}))$, i.e., $A_u^\ominus \cup A_a^\ominus = A \setminus B$, and $R_u^\ominus \cup R_a^\ominus \cup R_c^\ominus = R \setminus (\Im_{A \setminus B:B} \cup \Im_{A \setminus B})$.*

Proof.   Since $A_u^\oplus = A \setminus (afct(A, B, R \cup \Im_{A:B}) \cup afct(A, \Im_A, R))$ and $A_a^\oplus = afct(A, B, R \cup \Im_{A:B}) \cup afct(A, \Im_A, R) \cup B$, it holds that $A_u^\oplus \cup A_a^\oplus = A \cup B$; according to Lemma 7.1, $R_u^\oplus \cup R_a^\oplus \cup R_c^\oplus = R \cup \Im_{A:B} \cup \Im_A$. On the other hand, since $A_u^\ominus = A \setminus (A_a^\ominus \cup B)$, it holds that $A_u^\ominus \cup A_a^\ominus = (A \setminus (A_a^\ominus \cup B)) \cup A_a^\ominus = A \setminus B$; according to Lemma 7.2, $R_u^\ominus \cup R_a^\ominus \cup R_c^\ominus = R \setminus (\Im_{A \setminus B:B} \cup \Im_{A \setminus B})$.   □

Example 7.2.   Let $F = (A, R)$ be an argumentation framework, in which $A = \{a_1, a_2, a_3, a_4, a_5, a_6, a_7, a_8\}$ and $R = \{(a_1, a_2), (a_2, a_1), (a_1, a_3), (a_2, a_3), (a_3, a_4), (a_5, a_6), (a_6, a_5), (a_7, a_8), (a_8, a_7)\}$ (Figure 7.2). Let $(B, \Im_{A:B} \cup \Im_A)$ be an addition, in which $B = \{a_9, a_{10}\}$,

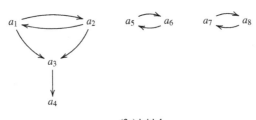

(3a) initial

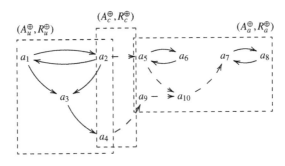

(3b) after addition

Figure 7.2   An example of the division of an updated argumentation framework.

$\mathfrak{I}_{A:B} = \{(a_4, a_9), (a_9, a_{10}), (a_5, a_{10}), (a_{10}, a_7)\}$, and $\mathfrak{I}_A = \{(a_2, a_5)\}$. According to Definition 7.7, the division of the updated argumentation framework $(A^\oplus, R^\oplus) = (A \cup B, R \cup (\mathfrak{I}_{A:B} \cup \mathfrak{I}_A))$ as well as the corresponding conditioned sub-framework are as follows:

$$(A_a^\oplus, R_a^\oplus) = (\{a_5, a_6, a_7, a_8, a_9, a_{10}\}, \{(a_5, a_6), (a_6, a_5), (a_7, a_8), (a_8, a_7),$$
$$(a_5, a_{10}), (a_9, a_{10}), (a_{10}, a_7)\})$$
$$(A_u^\oplus, R_u^\oplus) = (\{a_1, a_2, a_3, a_4\}, \{(a_1, a_2), (a_2, a_1), (a_1, a_3), (a_2, a_3), (a_3, a_4)\})$$
$$(A_c^\oplus, R_c^\oplus) = (\{a_2, a_4\}, \{(a_2, a_5), (a_4, a_9)\})$$
$$CSF_{add} = (A_a^\oplus \cup A_c^\oplus, R_a^\oplus \cup R_c^\oplus)$$

In this example, it is obvious that $(A^\oplus, R^\oplus)$ is equal to the combination of $(A_u^\oplus, R_u^\oplus)$ and CSF_{add}.

7.4 Computing the Semantics of an Updated Argumentation Framework Based on the Division

Based on the concept of the division of an argumentation framework, we are now ready to compute the semantics of the two kinds of sub-frameworks described above and combine them to form the semantics of the updated frameworks $(A^\oplus, R^\oplus)$ and $(A^\ominus, R^\ominus)$, respectively.

On the one hand, let $\mathscr{E}_\sigma(F)$ be the set of extensions of an argumentation framework $F = (A, R)$, under the argumentation semantics $\sigma \in \{adm, co, pr, gr\}$. According to the theory presented in Section 5.2, the set of extensions of the unaffected sub-framework $(A_u^\oplus, R_u^\oplus)$ with respect to an addition $(B, \mathfrak{I}_{A:B} \cup \mathfrak{I}_A)$ (respectively, $(A_u^\ominus, R_u^\ominus)$ with respect to a deletion $(B, \mathfrak{I}_{A \setminus B:B} \cup \mathfrak{I}_{A \setminus B})$) can be obtained directly:

$$\mathscr{E}_\sigma((A_u^\oplus, R_u^\oplus)) = \{E \cap A_u^\oplus \mid E \in \mathscr{E}_\sigma(F)\} \tag{7.7}$$

$$\mathscr{E}_\sigma((A_u^\ominus, R_u^\ominus)) = \{E \cap A_u^\ominus \mid E \in \mathscr{E}_\sigma(F)\} \tag{7.8}$$

On the other hand, according to Definitions 4.8 and 4.9, the extensions of conditioned sub-frameworks are not computed directly. They are related to the status of the conditioning arguments. In other words, we should firstly construct two sets of partially assigned sub-frameworks:

$$CSF_{add}[E_1] = (A_a^\oplus \cup A_c^\oplus, R_a^\oplus \cup R_c^\oplus)^{E_1}, \quad \forall E_1 \in \mathscr{E}_\sigma((A_u^\oplus, R_u^\oplus)) \tag{7.9}$$

$$CSF_{del}[E_1] = (A_a^\ominus \cup A_c^\ominus, R_a^\ominus \cup R_c^\ominus)^{E_1}, \quad \forall E_1 \in \mathscr{E}_\sigma((A_u^\ominus, R_u^\ominus)) \tag{7.10}$$

And then, the extensions of the partially assigned sub-frameworks, i.e., $\mathscr{E}_\sigma(CSF_{add}[E_1])$ and $\mathscr{E}_\sigma(CSF_{del}[E_1])$, can be obtained.

Based on the extensions of the two kinds of sub-frameworks, we obtain the extensions of $(A^{\oplus}, R^{\oplus})$ by combining $\mathscr{E}_\sigma((A_u^{\oplus}, R_u^{\oplus}))$ and $\mathscr{E}_\sigma(CSF_{add}[E_1])$, in which $E_1 \in \mathscr{E}_\sigma((A_u^{\oplus}, R_u^{\oplus}))$, and the extensions of $(A^{\ominus}, R^{\ominus})$ by combining $\mathscr{E}_\sigma((A_u^{\ominus}, R_u^{\ominus}))$ and $\mathscr{E}_\sigma(CSF_{del}[E_1])$, in which $E_1 \in \mathscr{E}_\sigma((A_u^{\ominus}, R_u^{\ominus}))$, in which $\sigma \in \{adm, co, pr, gr\}$.

$$CombExt_\sigma((A^{\oplus}, R^{\oplus})) = \{E_1 \cup E_2 \mid E_1 \in \mathscr{E}_\sigma((A_u^{\oplus}, R_u^{\oplus})) \wedge$$
$$E_2 \in \mathscr{E}_\sigma(CSF_{add}[E_1])\} \qquad (7.11)$$
$$CombExt_\sigma((A^{\ominus}, R^{\ominus})) = \{E_1 \cup E_2 \mid E_1 \in \mathscr{E}_\sigma((A_u^{\ominus}, R_u^{\ominus})) \wedge$$
$$E_2 \in \mathscr{E}_\sigma(CSF_{del}[E_1])\} \qquad (7.12)$$

7.5 An Illustrating Example

Now, let us consider the following example, which illustrates the process of computing the extensions of an updated argumentation framework based on the division. For simplicity and without loss of generality, we only discuss the case under preferred semantics.

Example 7.3. Continue Example 7.2. A set of extensions of F under preferred semantics are: $\mathscr{E}_{pr}(F) = \{E_{0,1}, E_{0,2}, E_{0,3}, E_{0,4}, E_{0,5}, E_{0,6}, E_{0,7}, E_{0,8}\}$, in which $E_{0,1} = \{a_1, a_4, a_5, a_7\}$, $E_{0,2} = \{a_1, a_4, a_5, a_8\}$, $E_{0,3} = \{a_1, a_4, a_6, a_7\}$, $E_{0,4} = \{a_1, a_4, a_6, a_8\}$, $E_{0,5} = \{a_2, a_4, a_5, a_7\}$, $E_{0,6} = \{a_2, a_4, a_5, a_8\}$, $E_{0,7} = \{a_2, a_4, a_6, a_7\}$, and $E_{0,8} = \{a_2, a_4, a_6, a_8\}$.

First, according to Formula 7.7, the extensions of $(A_u^{\oplus}, R_u^{\oplus})$ under preferred semantics are directly computed:

$$\mathscr{E}_{pr}((A_u^{\oplus}, R_u^{\oplus})) = \{E \cap A_u^{\oplus} \mid E \in \{E_{0,1}, E_{0,2}, \ldots, E_{0,8}\}\} = \{E_{1,1}, E_{1,2}\}, \text{ in which}$$
$$E_{1,1} = \{a_1, a_4\}, \text{ and}$$
$$E_{1,2} = \{a_2, a_4\}.$$

Second, get a set of partially assigned sub-frameworks according to conditioning arguments $A_c^{\oplus}$ with different statuses. It holds that $A_c^{\oplus}[E_{1,1}] = (\{a_4\}, \{a_2\}, \emptyset)$ and $A_c^{\oplus}[E_{1,2}] = (\{a_2, a_4\}, \emptyset, \emptyset)$. So, there are two different partially assigned sub-frameworks (Figure 7.3).

Third, compute the extensions of the partially assigned sub-frameworks:

$$\mathscr{E}_{pr}(CSF_{add}[E_{1,1}]) = \{E_{2,1}, E_{2,2}, E_{2,3}\}, \text{ in which } E_{2,1} = \{a_5, a_7\}, E_{2,2} = \{a_5, a_8\},$$
$$\text{and } E_{2,3} = \{a_6, a_8, a_{10}\};$$
$$\mathscr{E}_{pr}(CSF_{add}[E_{1,2}]) = \{E_{2,4}\}, \text{ in which } E_{2,4} = \{a_6, a_8, a_{10}\}.$$

Fourth, compute the results of combining the extensions of two kinds of sub-frameworks:

$$CombExt_{pr}((A^{\oplus}, R^{\oplus})) = \{E_1 \cup E_2 \mid E_1 \in \mathscr{E}_{pr}((A_u^{\oplus}, R_u^{\oplus})) \wedge$$
$$E_2 \in \mathscr{E}_{pr}(CSF_{add}[E_1])\}$$

$$= \{E_{1,1} \cup E_{2,1}, E_{1,1} \cup E_{2,2}, E_{1,1} \cup E_{2,3}, E_{1,2} \cup E_{2,4}\}$$
$$= \{\{a_1, a_4, a_5, a_7\}, \{a_1, a_4, a_5, a_8\}, \{a_1, a_4, a_6, a_8, a_{10}\},$$
$$\{a_2, a_4, a_6, a_8, a_{10}\}\}.$$

Here, we may verify that $E_{1,1} \cup E_{2,1}$, $E_{1,1} \cup E_{2,2}$, $E_{1,1} \cup E_{2,3}$, and $E_{1,2} \cup E_{2,4}$ are preferred extensions of $(A^\oplus, R^\oplus)$. Take $E_{1,1} \cup E_{2,3} = \{a_1, a_4, a_6, a_8, a_{10}\}$ for example. It holds that: (1) $E_{1,1} \cup E_{2,3} \subseteq A \cup B$. (2) $E_{1,1} \cup E_{2,3}$ is conflict-free. (3) Every argument in $E_{1,1} \cup E_{2,3}$ is acceptable with respect to $E_{1,1} \cup E_{2,3}$: a_1 is attacked by a_2, which is defeated by $a_1 \in E_{1,1} \subseteq E_{1,1} \cup E_{2,3}$; a_4 is attacked by a_3, which is defeated by $a_1 \in E_{1,1} \subseteq E_{1,1} \cup E_{2,3}$; a_6 is attacked by a_5, which is defeated by $a_6 \in E_{2,3} \subseteq E_{1,1} \cup E_{2,3}$; a_8 is attacked by a_7, which is defeated by $a_{10} \in E_{2,3} \subseteq E_{1,1} \cup E_{2,3}$; a_{10} is attacked by a_5 and a_9, in which a_5 is defeated by $a_6 \in E_{2,3} \subseteq E_{1,1} \cup E_{2,3}$ and a_9 is defeated by $a_4 \in A_c^\oplus \cap E_{1,1} = \{a_4\} \subseteq E_{1,1} \cup E_{2,3}$. (4) Every argument in $A \cup B$ that is acceptable with respect to $E_{1,1} \cup E_{2,3}$ is in $E_{1,1} \cup E_{2,3}$: every argument in $(A \cup B) \setminus (E_{1,1} \cup E_{2,3}) = \{a_2, a_3, a_5, a_7, a_9\}$ is not acceptable with respect to $E_{1,1} \cup E_{2,3}$. (5) $E_{1,1} \cup E_{2,3}$ is maximal.

On the other hand, if we directly compute the preferred extensions of $(A^\oplus, R^\oplus)$, then we have $\mathscr{E}_{pr}((A^\oplus, R^\oplus)) = \{E'_{0,1}, E'_{0,2}, E'_{0,3}, E'_{0,4}\} = \{\{a_1, a_4, a_5, a_7\}, \{a_1, a_4, a_5, a_8\}, \{a_1, a_4, a_6, a_8, a_{10}\}, \{a_2, a_4, a_6, a_8, a_{10}\}\}$. For every extension $E'_{0,i}$ $(i = 1, 2, 3, 4)$ in $\mathscr{E}_{pr}((A^\oplus, R^\oplus))$, there is a corresponding extension $E_{1,j} = E'_{0,i} \cap A_u^\oplus$ $(j = 1, 2)$ in $\mathscr{E}_{pr}((A_u^\oplus, R_u^\oplus))$ and a corresponding extension $E_{2,k} \cap A_a^\oplus$ $(k = 1, 2, 3, 4)$ in $\mathscr{E}_{pr}(CSF_{add}[E_{1,j}])$, such that $E'_{0,i} = E_{1,j} \cup E_{2,k}$. For example, as to $E'_{0,1} = \{a_1, a_4, a_5, a_7\}$, we have $E'_{0,1} \cap A_u^\oplus = \{a_1, a_4\} = E_{1,1}$ and $E'_{0,1} \cap A_a^\oplus = \{a_5, a_7\} = E_{2,1}$. It is obvious that $E'_{0,1} = E_{1,1} \cup E_{2,1}$.

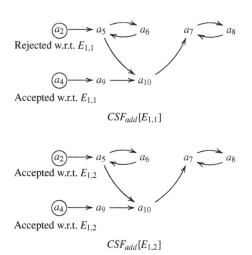

Figure 7.3 Two partially assigned sub-frameworks.

7.6 Conclusions

In this chapter, we have introduced a division-based method for computing the dynamic semantics of argumentation. It has the following characteristics:

1. *Generality*: It is a general theory in the sense of the following two aspects. First, this theory is applicable to some typical argumentation semantics, including admissible semantics, complete semantics, preferred semantics, and grounded semantics, etc. Second, this theory is able to treat with a general form of dynamics of argumentation, i.e., (i) the number of arguments and attacks to be added to or deleted from an argumentation system is unlimited; and (ii) both the addition and the deletion of arguments and/or attacks are applicable.
2. *Efficiency*: Qualitatively, it is obvious that in most cases (although not in all cases) the division-based method is more efficient. This is mainly due to the following two reasons. First, there exist linear time algorithms for the division of an argumentation framework (in that the problem of dividing an argumentation framework corresponds to finding the nodes reachable from a set of nodes in a directed graph). Second, when computing the extensions of a modified argumentation framework, we may reuse some previous computation, rather than simply recompute the status of each argument afresh.

With the above two characteristics, this theory is expected to be very useful in various kinds of argumentation-based systems, especially belief revision [4–7,9], deliberation [10–12], decision-making [14,15,17,18], and negotiation [19–23], within agents and multi-agent systems. The reason is that in these systems, underlying knowledge and information are often uncertain, incomplete, inconsistent and ever-changing. As a result, the corresponding argumentation systems are dynamic by nature. So, the efficient division-based method will facilitate the development of these systems.

In recent years, dynamics of argumentation has attracted some research efforts. However, up to now, its concept is still unclear. Different researchers treated it from different perspectives.

First, in the area of dialectical argumentation, the dynamics of argumentation means that the dialectical process of argumentation may change with the variations of knowledge at different stages [31–33]. This concept focuses on how a position (or a claim) can be proved with respect to a theory which can be revised dynamically. Researchers in this area did not care about the status evolution of the whole set of arguments within an argumentation framework, but only considered whether a specific position (or a claim) is acceptable according to a dialectical proof procedure where two parties (proposer and opposer) are involved.

Second, other researchers paid more attention to how the whole set of arguments and attacks of an argumentation system changes with the changing of underlying information, or how the status of arguments of an argumentation system evolves upon the changing of arguments and attacks. On the one hand, in [6], Capobianco et al used potential arguments and the instances of them to treat with the knowledge changing in dynamic environment. When perceptions change dynamically, the instances of potential arguments and attacks among them vary accordingly. As a result, the status of arguments is changed. Sharing some basic ideas with [6], Rotstein et al introduced the notion of dynamics into the concept of abstract argumentation frameworks, by including the concept of evidence [7]. They proposed a concept of *Dynamic Argumentation Framework* (corresponding to potential arguments in [6]), from which a static instance can be obtained, according to a varying set of evidence (corresponding to perceptions in [6]). However, both of them have not considered how to dynamically compute the status of arguments without computing the whole set of arguments in each time point. On the other hand, some researchers have performed some works towards this issue. For example, Moguillansky et al studied the dynamics of argumentation based on the idea of classical belief revision, i.e., when an argument is added to a system, the revision operator will change the set of arguments, with an objective that the newly added argument is accepted [8]. Boella et al studied the dynamics of argumentation by exploring the principles where the extension does not change when an argument or an attack between them is changed, but without considering how to dynamically compute the status of arguments when the status of existing arguments is affected [27,34]. Baroni et al proposed a notion of directionality, which says that unattacked sets are unaffected by the remaining part of the argumentation framework as far as extensions are concerned [29,30], but without considering the dynamics of argumentation systems. Cayrol et al explored the impact of the arrival of a new argument on the outcome of an argumentation framework, by defining a typology of refinement (i.e. adding an argument), and defining principles and condition so that each type of refinement becomes a classical revision [28]. However, they did not focus on how to compute the status of arguments when an argumentation framework is expanded with a set of arguments or attacks. Amgoud and Vesic studied how the acceptability of arguments evolves when the decision system is extended by new arguments without computing the whole extensions [16]. Their theory is focused on the revision of a particular argument (a practical argument or an epistemic argument). On the contrary, we are more interested in the issue of how the status of all arguments in an argumentation framework changes when any variations arise.

Based on the above analysis, we may conclude that the division-based method is in line with [16,27–29], etc., but has the two characteristics (*generality* and *efficiency*) mentioned above.

References

[1] A. Kakas, R. Miller, F. Toni, An argumentation framework for reasoning about actions and change, in: Proceedings of the 5th International Conference on Logic Programming and Nonmonotonic Reasoning, 1999, pp. 78–91.

[2] T.J.M. Bench-Capon, P.E. Dunne, Argumentation in artificial intelligence, Artificial Intelligence 171 (10–15) (2007) 619–641.

[3] G. Boella, G. Pigozzi, L. van der Torre, Normative framework for normative system change, in: Proceedings of the 8th International Conference on Autonomous Agents and Multiagent Systems, 2009, pp. 169–176.

[4] M.A. Falappa, G. Kern-Isberner, G.R. Simari, Explanations, belief revision and defeasible reasoning, Artificial Intelligence 141 (1–2) (2002) 1–28.

[5] M. Capobianco, C.I. Chesñevar, G.R. Simari, An argument-based framework to model an agent's beliefs in a dynamic environment, in: Proceedings of the 1st International Workshop on Argumentation in Multi-Agent Systems, 2005, pp. 95–110.

[6] M. Capobianco, C.I. Chesñevar, G.R. Simari, Argumentation and the dynamics of warranted beliefs in changing environments, Journal of Autonomous Agents and Multi-Agent Systems 11 (2) (2005) 127–151.

[7] N.D. Rotstein, M.O. Moguillansky, A.J. García, G.R. Simari, An abstract argumentation framework for handling dynamics, in: Proceedings of the 12th International Workshop on Non-Monotonic Reasoning, 2008, pp. 131–139.

[8] M.O. Moguillansky, N.D. Rotstein, M.A. Falappa, A.J. García, G.R. Simari, Argument theory change applied to defeasible logic programming, in: Proceedings of the Twenty-Third AAAI Conference on Artificial Intelligence, 2008, pp. 132–137.

[9] M.A. Falappa, G. Kern-Isberner, G.R. Simari, Belief revision and argumentation theory, in: Argumentation in Artificial Intelligence, 2009, pp. 341–360.

[10] L. Amgoud, A formal framework for handling conflicting desires, in: Proceedings of the 7th European Conference on Symbolic and Quantitative Approaches to Reasoning with Uncertainty, 2003, pp. 552–563.

[11] S. Modgil, M. Luck, Argumentation based resolution of conflicts between desires and normative goals, in: Proceedings of the 5th International Workshop on Argumentation in Multi-Agent Systems, 2009, pp. 19–36.

[12] D. Gaertner, J.A. Rodríguez-Aguilar, F. Toni, Agreeing on institutional goals for multi-agent societies, in: Proceedings of the COIN@AAAI'08 Workshop, 2009, pp. 1–16.

[13] B. Liao, H. Huang, An argumentation-based flexible agent with dynamic rules of inference, in: Proceedings of the 21st International Conference on Tools with Artificial Intelligence, 2009, pp. 284–291.

[14] A.C. Kakas, P. Moraïtis, Argumentation based decision making for autonomous agents, in: Proceedings the 2nd International Joint Conference on Autonomous Agents and Multiagent Systems, 2003, pp. 883–890.

[15] J. Fox, D. Glasspool, D. Grecu, S. Modgil, M. South, V. Patkar, Argumentation-based inference and decision making-medical perspective, IEEE Intelligent Systems 22 (6) (2007) 34–41.

[16] L. Amgoud, S. Vesic, On revising argumentation-based decision systems, in: Proceedings of the 10th European Conference on Symbolic and Quantitative Approaches to Reasoning with Uncertainty, 2009, pp. 71–82.

[17] L. Amgoud, H. Prade, Using arguments for making and explaining decisions, Artificial Intelligence 173 (3–4) (2009) 413–436.

[18] P.-A. Matt, F. Toni, J. Vaccari, Dominant decisions by argumentation agents, in: Proceedings of the 6th International Workshop on Argumentation in Multi-Agent Systems, 2010, pp. 42–59.

[19] S. Parsons, C.A. Sierra, N.R. Jennings, Agents that reason and negotiate by arguing, Journal of Logic and Computation 8 (3) (1998) 261–292.

[20] S. Kraus, K. Sycara, A. Evenchik, Reaching agreements through argumentation: a logical model and implementation, Artificial Intelligence 104 (1–2) (1998) 1–69.

[21] L. Amgoud, S. Parsons, N. Maudet, Argument, dialogue and negotiation, in: Proceedings of the 14th European Conference on Artificial Intelligence, 2000, pp. 338–342.

[22] I. Rahwan, S.D. Ramchurn, N.R. Jennings, P. McBurney, S. Parsons, L. Sonenberg, Argumentation-based negotiation, Knowledge Engineering Review 18 (4) (2003) 343–375.

[23] L. Amgoud, Y. Dimopoulos, P. Moraitis, A general framework for argumentation-based negotiation, in: Proceedings of the 4th International Workshop on Argumentation in Multi-Agent Systems, 2008, pp. 1–17.

[24] M.A. Falappa, A.J. García, G.R. Simari, Belief dynamics and defeasible argumentation in rational agents, in: Proceedings of the 10th International Workshop on Non-Monotonic Reasoning, 2004, pp. 164–170.

[25] B. Liao, H. Huang, ANGLE: an autonomous, normative and guidable agent with changing knowledge, Information Sciences 180 (17) (2010) 3117–3139.

[26] S. Coste-Marquis, C. Devred, S. Konieczny, M. Lagasquie-Schiex, P. Marquis, On the merging of Dung's argumentation systems, Artificial Intelligence 171 (10–15) (2007) 730–753.

[27] G. Boella, S. Kaci, L. van der Torre, Dynamics in argumentation with single extensions: abstraction principles and the grounded extension, in: Proceedings of the 10th European Conference on Symbolic and Quantitative Approaches to Reasoning with Uncertainty, 2009, pp. 107–118.

[28] C. Cayrol, F.D. de St-Cyr, M. Lagasquie-Schiex, Revision of an argumentation system, in: Proceedings of the 11th International Conference on the Principles of Knowledge Representation and Reasoning, 2008, pp. 124–134.

[29] P. Baroni, M. Giacomin, G. Guida, SCC-recursiveness: a general schema for argumentation semantics, Artificial Intelligence 168 (1–2) (2005) 162–210.

[30] P. Baroni, M. Giacomin, On principle-based evaluation of extension-based argumentation semantics, Artificial Intelligence 171 (10–15) (2007) 675–700.

[31] D.V. Carbogim, Dynamics on Formal Argumentation, Ph.D Thesis, University of Edinburgh, 2000.

[32] H. Prakken, Relating protocols for dynamic dispute with logics for defeasible argumentation, Synthese 127 (1–2) (2001) 187–219.

[33] K. Okuno, K. Takahashi, Argumentation system with changes of an agent's knowledge base, in: Proceedings of the Twenty-first International Joint Conference on Artificial Intelligence, 2009, pp. 226–232.

[34] G. Boella, S. Kaci, L. van der Torre, Dynamics in argumentation with single extensions: attack refinement and the grounded extension, in: Proceedings of the 8th International Conference on Autonomous Agents and Multiagent Systems, 2009, pp. 1213–1214.

CHAPTER 8

An Approach for Partial Semantics of Argumentation

Chapter Outline

8.1 Introduction

When querying the status of some arguments in an abstract argumentation framework, we may only take into consideration a subset of arguments in the argumentation framework that are *relevant* to these arguments. This phenomenon appears in many situations. Let us consider the following two examples.

First, when developing proof theories and algorithms for argumentation frameworks [1], for a given semantics, there are some "local" questions concerning the existence of extensions with respect to a subset B of arguments, such as:

(a) Is B contained in an extension? (credulous membership question)
(b) Is B contained in all extensions? (skeptical membership question)
(c) Is B attacked by an extension?
(d) Is B attacked by all extensions?

When answering these "local" questions, we might not have to figure out the extensions of a whole framework, but the extensions of a part of the framework.

Efficient Computation of Argumentation Semantics. http://dx.doi.org/10.1016/B978-0-12-410406-8.00008-7

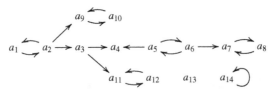

Figure 8.1 Argumentation framework $F_{8.1}$.

Example 8.1. Let $F_{8.1}$ be an abstract argumentation framework (Figure 8.1). Given $B_1 = \{a_9, a_{10}\}$, according to the directionality of argumentation semantics, only the status of a_1 and a_2 may affect the status of a_9 and a_{10}. In other words, when computing the status of arguments in B_1, we may only consider the set $\{a_1, a_2, a_9, a_{10}\}$ of arguments.

Second, when agents engage in dialogues, they have goals to make some arguments (say, a set B of arguments) acceptable or unacceptable [2]. At each turn, an agent may choose some arguments from a set of available arguments, and add them to the framework. For each addition, the status of arguments of the framework may change accordingly. However, since only the status evolution of arguments in B is necessary to be figured out, other arguments that are irrelevant to B may not be taken into consideration.

Example 8.2. For the $F_{8.1}$ in Example 8.1, suppose that an agent wants to make a_9 skeptically accepted under preferred semantics, and she owns a set of arguments $\{b_1, b_2, b_3\}$ and the set of associated attacks $\{(b_1, a_2), (b_2, a_4), (b_3, a_{10})\}$. Since only the status of b_1 and b_3 may affect the status of a_9, the agent may not consider b_2, which has no effect on the status of a_9. Meanwhile, when b_1, b_3 and the associated attacks (b_1, a_2) and (b_3, a_{10}) are added to $F_{8.1}$, although the status of arguments a_3, a_4, a_{11} and a_{12} is also affected by the addition, they may not be involved in the computation, in that they have no effect on the status of a_9 (and therefore are irrelevant to a_9).

Inspired by the above phenomenon, we propose a general approach to efficiently evaluate the status of a part of arguments in an argumentation framework. The basic idea of this method is as follows.

Given an argumentation framework and a subset B of arguments within it, we firstly identify the *minimal* set of arguments that are *relevant* to the arguments in B (called the set of *relevant arguments* of B). Then, under a semantics under which the mappings between global semantics and local semantics are sound and complete, the set of extensions of the sub-framework that is induced by the set of relevant arguments of B (called a *partial semantics* of the argumentation framework with respect to B) can be evaluated locally.

8.2 The Definition of Partial Semantics of Argumentation

As illustrated in Example 8.1, given $F = (A, R)$ and $B \subseteq A$, only those arguments, each of which is in B or has a path to the arguments in B, might be relevant to the status of arguments in B. For convenience, we call them the set of *relevant arguments* of B, denoted as $rlvt_F(B)$.

Definition 8.1.   Given $F = (A, R)$ and $B \subseteq A$, the set of relevant arguments of B is defined as follows:

$$rlvt_F(B) = B \cup$$

$$\left(\bigcup_{\alpha \in B} \{\beta \in A \setminus B : \text{there is a path from } \beta \text{ to } \alpha \text{ with respect to } R\} \right)$$

Example 8.3.   Continue Example 8.1. Given $B_1 = \{a_9, a_{10}\}$, according to Definition 8.1, $rlvt_{F_{8.1}}(B_1) = \{a_1, a_2, a_9, a_{10}\}$. In addition, given $B_2 = \{a_{13}\}$ and $B_3 = \{a_{14}\}$, we have $rlvt_{F_{8.1}}(B_2) = \{a_{13}\}$, and $rlvt_{F_{8.1}}(B_3) = \{a_{14}\}$.

Based on Definition 8.1, some basic properties of the set of relevant arguments of a given subset are formulated in the following proposition.

Proposition 8.1.   *Let $F = (A, R)$ be an abstract argumentation framework, and $B, C \subseteq A$ sets of arguments. It holds that:*

(1) $rlvt_F(B) \supseteq B$;
(2) $rlvt_F(B)^- = \emptyset$, *i.e., $rlvt_F(B)$ is an unattacked set;*
(3)   *if $B \supseteq C$, then $rlvt_F(B) \supseteq rlvt_F(C)$;*
(4) $rlvt_F(B \cup C) = rlvt_F(B) \cup rlvt_F(C)$.

Proof.

(1)   Obvious.
(2)   Assume the contrary. Then, $\exists \gamma \in (A \setminus rlvt_F(B))$, such that γ attacks an argument α in $rlvt_F(B)$. It follows that there is a path from γ to α. If $\alpha \in B$, according to Definition 8.1, $\gamma \in rlvt_F(B)$, contradicting "$\gamma \in (A \setminus rlvt_F(B))$" (i.e., $\gamma \notin rlvt_F(B)$). Otherwise, if $\alpha \in rlvt_F(B) \setminus B$, then: according to Definition 8.1, $\exists \alpha' \in B$, such that there is a path from α to α'. It follows that there is a path from γ to $\alpha' \in B$, and therefore $\gamma \in rlvt_F(B)$, contradicting "$\gamma \in (A \setminus rlvt_F(B))$".
(3)   Given $B \supseteq C$, according to Definition 8.1, it is obvious that $rlvt_F(B) \supseteq rlvt_F(C)$.
(4)   On the one hand, since $rlvt_F(B \cup C) \supseteq rlvt_F(B)$ and $rlvt_F(B \cup C) \supseteq rlvt_F(C)$, it holds that $rlvt_F(B \cup C) \supseteq rlvt_F(B) \cup rlvt_F(C)$.

On the other hand, $\forall \alpha \in rlvt_F(B \cup C)$, according to Definition 8.1, there are two possible cases:

(a) $\alpha \in B \cup C$. In this case, since $B \cup C \subseteq rlvt_F(B) \cup rlvt_F(C)$, it holds that $\alpha \in rlvt_F(B) \cup rlvt_F(C)$.
(b) $\exists \beta \in B \cup C$, such that there is a path from α to β. In this case, $\alpha \in rlvt_F(B)$ (when $\beta \in B$) or $\alpha \in rlvt_F(C)$ (when $\beta \in C$), i.e., $\alpha \in rlvt_F(B) \cup rlvt_F(C)$.

In both cases, it holds that $\forall \alpha \in rlvt_F(B \cup C)$, $\alpha \in rlvt_F(B) \cup rlvt_F(C)$. It follows that $rlvt_F(B \cup C) \subseteq rlvt_F(B) \cup rlvt_F(C)$.

Since $rlvt_F(B \cup C) \supseteq rlvt_F(B) \cup rlvt_F(C)$ and $rlvt_F(B \cup C) \subseteq rlvt_F(B) \cup rlvt_F(C)$, we have $rlvt_F(B \cup C) = rlvt_F(B) \cup rlvt_F(C)$.

Since $rlvt_F(B)$ is an unattacked set, we may use $rlvt_F(B)$ to induce an unconditioned sub-framework of $F : (rlvt_F(B), R_{rlvt_F(B)})$, in which $R_{rlvt_F(B)} = R \cap (rlvt_F(B) \times rlvt_F(B))$. Under a semantics $\sigma \in \{adm, co, pr, gr\}$, the extensions of $(rlvt_F(B), R_{rlvt_F(B)})$ can be computed locally. We call $\mathcal{E}_\sigma((rlvt_F(B), R_{rlvt_F(B)}))$ a *partial semantics* of F with respect to B.

Definition 8.2. Let $F = (A, R)$ be an abstract argumentation framework, $B \subseteq A$ a set of arguments, and $rlvt_F(B)$ the set of relevant arguments of B. Under a semantics $\sigma \in \{adm, co, pr, gr\}$, the partial semantics of F with respect to B is defined as $\mathcal{E}_\sigma((rlvt_F(B), R_{rlvt_F(B)}))$.

Proposition 8.2. *Let $F = (A, R)$ be an abstract argumentation framework, and $B \subseteq A$ a set of arguments. Under a semantics $\sigma \in \{adm, co, pr, gr\}$, for every argument $\alpha \in B$, α is skeptically justified (respectively, credulously justified and indefensible) with respect to $\mathcal{E}_\sigma(F)$, if and only if α is skeptically justified (respectively, credulously justified and indefensible) with respect to $\mathcal{E}_\sigma((rlvt_F(B), R_{rlvt_F(B)}))$.*

Proof. $(\Rightarrow)$

- If α is skeptically justified with respect to $\mathcal{E}_\sigma(F)$, then $\forall E \in \mathcal{E}_\sigma(F)$, $\alpha \in E$. Assume that $\exists E_1 \in \mathcal{E}_\sigma((rlvt_F(B), R_{rlvt_F(B)}))$, such that $\alpha \notin E_1$. Since $rlvt_F(B)^- = \emptyset$, according to the theories presented in Chapter 5, $\mathcal{E}_\sigma((rlvt_F(B), R_{rlvt_F(B)})) = \{E \cap rlvt_F(B) \mid E \in \mathcal{E}_\sigma(F)\}$. It follows that $\exists E \in \mathcal{E}_\sigma(F)$, such that $E_1 = E \cap rlvt_F(B)$. Since $\alpha \notin E_1$ and $\alpha \in rlvt_F(B)$, it holds that $\alpha \notin E$, contradicting "$\forall E \in \mathcal{E}_\sigma(F)$, $\alpha \in E$". Therefore, $\forall E_1 \in \mathcal{E}_\sigma((rlvt_F(B), R_{rlvt_F(B)}))$, $\alpha \in E_1$, i.e., α is skeptically justified with respect to $\mathcal{E}_\sigma((rlvt_F(B), R_{rlvt_F(B)}))$.
- If α is credulously justified with respect to $\mathcal{E}_\sigma(F)$, then $\exists E \in \mathcal{E}_\sigma(F)$, such that $\alpha \in E$. According to the theories presented in Chapter 5, $E \cap rlvt_F(B) \in \mathcal{E}_\sigma((rlvt_F(B), R_{rlvt_F(B)}))$. Since $\alpha \in rlvt_F(B)$ and $\alpha \in E$, it holds that $\alpha \in E \cap rlvt_F(B)$. In other words, there exists an extension in $\mathcal{E}_\sigma((rlvt_F(B), R_{rlvt_F(B)}))$, such that α is in this extension, i.e., α is credulously justified with respect to $\mathcal{E}_\sigma((rlvt_F(B), R_{rlvt_F(B)}))$.

- If α is indefensible with respect to $\mathscr{E}_\sigma(F)$, then $\forall E \in \mathscr{E}_\sigma(F)$, $\alpha \notin E$. Assume that $\exists E_1 \in \mathscr{E}_\sigma((rlvt_F(B), R_{rlvt_F(B)}))$, such that $\alpha \in E_1$. As mentioned in the first item, $\mathscr{E}_\sigma((rlvt_F(B), R_{rlvt_F(B)})) = \{E \cap rlvt_F(B) \mid E \in \mathscr{E}_\sigma(F)\}$. It follows that $\exists E \in \mathscr{E}_\sigma(F)$, such that $E_1 = E \cap rlvt_F(B)$. Since $\alpha \in E_1$, it holds that $\alpha \in E$, contradicting "$\forall E \in \mathscr{E}_\sigma(F)$, $\alpha \notin E$". Therefore, $\forall E_1 \in \mathscr{E}_\sigma((rlvt_F(B), R_{rlvt_F(B)}))$, $\alpha \notin E_1$, i.e., α is indefeasible with respect to $\mathscr{E}_\sigma((rlvt_F(B), R_{rlvt_F(B)}))$.

($\Leftarrow$)

- If α is skeptically justified with respect to $\mathscr{E}_\sigma((rlvt_F(B), R_{rlvt_F(B)}))$, then for all $E_1 \in \mathscr{E}_\sigma((rlvt_F(B), R_{rlvt_F(B)}))$, $\alpha \in E_1$. Assume that $\exists E \in \mathscr{E}_\sigma(F)$, such that $\alpha \notin E$. According to the theories presented in Chapter 5, $E \cap rlvt_F(B) \in \mathscr{E}_\sigma((rlvt_F(B), R_{rlvt_F(B)}))$. Since $\alpha \notin E$, it holds that $\alpha \notin E \cap rlvt_F(B)$, contradicting "$\forall E_1 \in \mathscr{E}_\sigma((rlvt_F(B), R_{rlvt_F(B)}))$, $\alpha \in E_1$". Therefore, $\forall E \in \mathscr{E}_\sigma(F)$, $\alpha \in E$, i.e., α is skeptically justified with respect to $\mathscr{E}_\sigma(F)$.
- If α is credulously justified with respect to $\mathscr{E}_\sigma((rlvt_F(B), R_{rlvt_F(B)}))$, then there exists $E_1 \in \mathscr{E}_\sigma((rlvt_F(B), R_{rlvt_F(B)}))$, such that $\alpha \in E_1$. Assume that $\nexists E \in \mathscr{E}_\sigma(F)$, such that $\alpha \in E$. Since $rlvt_F(B)^- = \emptyset$, according to the theories presented in Chapter 5, $\mathscr{E}_\sigma((rlvt_F(B), R_{rlvt_F(B)})) = \{E \cap rlvt_F(B) \mid E \in \mathscr{E}_\sigma(F)\}$. It follows that $\exists E \in \mathscr{E}_\sigma(F)$, such that $E_1 = E \cap rlvt_F(B)$. Since $\alpha \in E_1$, it holds that $\alpha \in E$, contradicting "$\nexists E \in \mathscr{E}_\sigma(F)$, $\alpha \in E$". Therefore, $\exists E \in \mathscr{E}_\sigma(F)$, such that $\alpha \in E$, i.e., α is credulously justified with respect to $\mathscr{E}_\sigma(F)$.
- If α is indefeasible with respect to $\mathscr{E}_\sigma((rlvt_F(B), R_{rlvt_F(B)}))$, then for all $E_1 \in \mathscr{E}_\sigma((rlvt_F(B), R_{rlvt_F(B)}))$, $\alpha \notin E_1$. Assume that $\exists E \in \mathscr{E}_\sigma(F)$, such that $\alpha \in E$. According to the theories presented in Chapter 5, $E \cap rlvt_F(B) \in \mathscr{E}_\sigma((rlvt_F(B), R_{rlvt_F(B)}))$. Since $\alpha \in E$ and $\alpha \in rlvt_F(B)$, it holds that $\alpha \in E \cap rlvt_F(B)$, contradicting "$\forall E_1 \in \mathscr{E}_\sigma((rlvt_F(B), R_{rlvt_F(B)}))$, $\alpha \notin E_1$". Therefore, $\forall E \in \mathscr{E}_\sigma(F)$, $\alpha \notin E$, i.e., α is indefeasible with respect to $\mathscr{E}_\sigma(F)$.

According to this proposition, when querying the justification status of some arguments in an argumentation framework, we only need to compute the partial semantics of the argumentation framework with respect to these arguments.

Example 8.4. Continue Example 8.3. Let us consider the following two approaches.

Firstly, use the partial semantics of $F_{8.1}$ to evaluate the status of arguments in B_1, B_2 and B_3, respectively.

Under preferred semantics,

$$\mathscr{E}_{pr}((rlvt_{F_{8.1}}(B_1), R_{rlvt_{F_{8.1}}(B_1)})) = \{E_{1,1}, E_{1,2}, E_{1,3}\}, \text{ in which}$$
$$E_{1,1} = \{a_1, a_9\},$$
$$E_{1,2} = \{a_1, a_{10}\},$$

$$E_{1,3} = \{a_2, a_{10}\};$$
$$\mathscr{E}_{pr}((rlvt_{F_{8.1}}(B_2), R_{rlvt_{F_{8.1}}}(B_2))) = \{E_{2,1}\}, \text{ in which}$$
$$E_{2,1} = \{a_{13}\};$$
$$\mathscr{E}_{pr}((rlvt_{F_{8.1}}(B_2), R_{rlvt_{F_{8.1}}}(B_3))) = \{E_{3,1}\}, \text{ in which}$$
$$E_{3,1} = \emptyset.$$

According to these partial semantics, it holds that a_9 and a_{10} are credulously justified with respect to $\mathscr{E}_{pr}((rlvt_{F_{8.1}}(B_1), R_{rlvt_{F_{8.1}}}(B_1)))$, a_{13} is skeptically justified with respect to $\mathscr{E}_{pr}((rlvt_{F_{8.1}}(B_2), R_{rlvt_{F_{8.1}}}(B_2)))$, and a_{14} is indefensible with respect to $\mathscr{E}_{pr}((rlvt_{F_{8.1}}(B_2), R_{rlvt_{F_{8.1}}}(B_3)))$.

Secondly, use the (whole) semantics of $F_{8.1}$ to evaluate the status of the above-mentioned arguments. Under preferred semantics,

$$\mathscr{E}_{pr}(F_{8.1}) = \{E_{0,1}, E_{0,2}, \ldots, E_{0,12}\}, \text{ in which}$$
$$E_{0,1} = \{a_1, a_3, a_5, a_7, a_9, a_{12}, a_{13}\},$$
$$E_{0,2} = \{a_1, a_3, a_5, a_8, a_9, a_{12}, a_{13}\},$$
$$E_{0,3} = \{a_1, a_3, a_6, a_8, a_9, a_{12}, a_{13}\};$$
$$E_{0,4} = \{a_1, a_3, a_5, a_7, a_{10}, a_{12}, a_{13}\},$$
$$E_{0,5} = \{a_1, a_3, a_5, a_8, a_{10}, a_{12}, a_{13}\},$$
$$E_{0,6} = \{a_1, a_3, a_6, a_8, a_{10}, a_{12}, a_{13}\}.$$
$$E_{0,7} = \{a_2, a_5, a_7, a_{10}, a_{11}, a_{13}\},$$
$$E_{0,8} = \{a_2, a_5, a_8, a_{10}, a_{11}, a_{13}\},$$
$$E_{0,9} = \{a_2, a_4, a_6, a_8, a_{10}, a_{11}, a_{13}\};$$
$$E_{0,10} = \{a_2, a_5, a_7, a_{10}, a_{12}, a_{13}\},$$
$$E_{0,11} = \{a_2, a_5, a_8, a_{10}, a_{12}, a_{13}\},$$
$$E_{0,12} = \{a_2, a_4, a_6, a_8, a_{10}, a_{12}, a_{13}\}.$$

It holds that a_9 and a_{10} are credulously justified, a_{13} is skeptically justified, and a_{14} is indefensible, with respect to $\mathscr{E}_{pr}(F_{8.1})$. According to Proposition 8.2, the results of these two approaches are the same. However, the latter is obviously more complex.

8.3 Basic Properties of Partial Semantics of Argumentation

According to the previous section, given $F = (A, R)$ and $B \subseteq A$, we may compute the partial semantics of F with respect to B independently, without computing the status of other arguments that are irrelevant to B. Furthermore, it is desirable that after the partial semantics of F with respect to some sets of arguments obtained, they can be reused in the subsequent computation. This vision is embodied by the following three basic properties of partial semantics of argumentation: *monotonicity*, *combinability* and *extensibility*.

8.3.1 Monotonicity of Partial Semantics

Given $F = (A, R)$ and $B, B' \subseteq A$, the *monotonicity* of partial semantics of argumentation can be informally expressed as:

> If $rlvt_F(B) \subseteq rlvt_F(B')$, then the justification status of each argument in B evaluated with respect to $\mathscr{E}_\sigma((rlvt_F(B), R_{rlvt_F(B)}))$ is the same as the one evaluated with respect to $\mathscr{E}_\sigma((rlvt_F(B'), R_{rlvt_F(B')}))$.

Since $rlvt_F(B) \subseteq rlvt_F(B')$ and $rlvt_F(B)$ is an unattacked set (according to Proposition 8.1), $(rlvt_F(B), R_{rlvt_F(B)})$ is an unconditioned sub-framework of $(rlvt_F(B'), R_{rlvt_F(B')})$. Given a semantics σ under which there exist sound and complete mappings between global semantics and local semantics, $\mathscr{E}_\sigma((rlvt_F(B), R_{rlvt_F(B)}))$ can be regarded as a partial semantics of $(rlvt_F(B'), R_{rlvt_F(B')})$ with respect to B. According to Proposition 8.2, we directly have the following corollary.

Corollary 8.1. *Given $F = (A, R)$ and $B, B' \subseteq A$, if $rlvt_F(B) \subseteq rlvt_F(B')$, then: under a semantics $\sigma \in \{adm, co, pr, gr\}$, for every argument $\alpha \in B$, α is skeptically justified (respectively, credulously justified and indefensible) with respect to $\mathscr{E}_\sigma((rlvt_F(B'), R_{rlvt_F(B')}))$, if and only if α is skeptically justified (respectively, credulously justified and indefensible) with respect to $\mathscr{E}_\sigma((rlvt_F(B), R_{rlvt_F(B)}))$.*

The above property indicates that after we get the partial semantics of F with respect to B' (i.e., $\mathscr{E}_\sigma((rlvt_F(B'), R_{rlvt_F(B')}))$), we may evaluate the justification status of arguments in B with respect to $\mathscr{E}_\sigma((rlvt_F(B'), R_{rlvt_F(B')}))$ directly, and do not need to compute the partial semantics of F with respect to B.

8.3.2 Extensibility of Partial Semantics

Given $F = (A, R)$ and $B, B' \subseteq A$, the *extensibility* of partial semantics of argumentation can be informally expressed as follows:

> If $rlvt_F(B) \subseteq rlvt_F(B')$, then the partial semantics of F with respect to B can be *extended* to form the partial semantics of F with respect to B'.

Let $Comp = rlvt_F(B') \setminus rlvt_F(B)$ denote the *complement* of $rlvt_F(B)$ in $rlvt_F(B')$. Given that the partial semantics of F with respect to B has been obtained, is it feasible to first compute

(a)   (b)

Figure 8.2 Two sub-frameworks of $F_{8.1}$: the sub-framework induced by B_4 (a) is an unconditioned sub-framework, whose extensions can be computed according to Definition 2.2, while the sub-framework induced by $Comp_1$ (b) is a conditioned sub-framework, which is conditioned by some arguments outside the sub-framework.

the partial semantics of F with respect to $Comp$, and then combine it with the former to form the partial semantics of F with respect to B'?

As we know, under a semantics $\sigma \in \{adm, co, pr, gr\}$, if $Comp$ is an unattacked set, then the partial semantics of F with respect to $Comp$ can be computed independently. However, this condition might not hold in many cases. Let us consider the following example.

Example 8.5.   With respect to $F_{8.1}$ in Figure 8.1, let $B_4 = \{a_3\}$ and $B_5 = \{a_{11}, a_{12}\}$. It follows that $rlvt_{F_{8.1}}(B_4) = \{a_1, a_2, a_3\}$ and $rlvt_{F_{8.1}}(B_5) = \{a_1, a_2, a_3, a_{11}, a_{12}\}$. Let $Comp_1 = rlvt_F(B_5) \setminus rlvt_F(B_4)$. We have $Comp_1 = \{a_{11}, a_{12}\}$, and $Comp_1^- = \{a_3\} \neq \emptyset$. So, $Comp_1$ is not an unattacked set, and the sub-framework induced by $Comp_1$ is a conditioned sub-framework (as shown in Figure 8.2(b)). In other words, since the status of arguments in $Comp_1$ is related to the status of some arguments outside $Comp_1$ (i.e., a_3), the extensions of the sub-framework induced by $Comp_1$ can not be computed independently.

Then, we may first construct a conditioned sub-framework for $Comp$. Since $Comp$ is contained in $F = (A, R)$ (in other words, $Comp \subseteq A$), the set of conditioning arguments of $Comp$ is equal to $Comp^-$. So, the conditioned sub-framework for $Comp$ is $(Comp \cup Comp^-, R_{Comp} \cup I_{Comp})$, in which $R_{Comp} = R \cap (Comp \times Comp)$ and $I_{Comp} = R \cap (Comp^- \times Comp)$.

Then, we need to prove that $Comp^-$ is contained in $rlvt_F(B)$, so that the status of arguments in $Comp^-$ can be assigned according to the extensions of $(rlvt_F(B), R_{rlvt_F(B)})$. For this purpose, we have the following proposition.

Proposition 8.3.   *Let $F = (A, R)$ be an abstract argumentation framework, $B, B' \subseteq A$ sets of arguments, and $Comp = rlvt_F(B') \setminus rlvt_F(B)$ the complement of $rlvt_F(B)$ in $rlvt_F(B')$. If $rlvt_F(B) \subseteq rlvt_F(B')$, then $Comp^- \subseteq rlvt_F(B)$.*

Proof.   Assume that $\exists \alpha \in Comp^- \subseteq A$, such that $\alpha \notin rlvt_F(B)$. Since $\alpha \in Comp^-$ and $Comp = rlvt_F(B') \setminus rlvt_F(B)$, according to Formula 2.1, it holds that $\alpha \notin rlvt_F(B') \setminus rlvt_F(B)$, and $\exists \beta \in Comp$, such that $(\alpha, \beta) \in R$.

Since $\alpha \notin rlvt_F(B') \setminus rlvt_F(B)$, and $\alpha \notin rlvt_F(B)$, it holds that $\alpha \notin rlvt_F(B')$, and therefore $\alpha \notin B' \subseteq rlvt_F(B')$.

Since $\beta \in Comp$ and $Comp = rlvt_F(B') \setminus rlvt_F(B)$, it holds that $\beta \in rlvt_F(B')$ and $\beta \notin rlvt_F(B)$. By $\beta \in rlvt_F(B')$, we have the following two possible cases:

(a) $\beta \in B'$. In this case, since $(\alpha, \beta) \in R$ (there is a path from α to β with respect to R) and $\alpha \notin B'$, according to Definition 8.1, $\alpha \in rlvt_F(B')$, contradicting $\alpha \notin rlvt_F(B')$.

(b) $\beta \in rlvt_F(B') \setminus B'$. In this case, $\exists \gamma \in B'$, such that there is a path from β to γ. It follows that there is a path from α to γ. According to Definition 8.1, $\alpha \in rlvt_F(B')$, contradicting $\alpha \notin rlvt_F(B')$.

So, $\forall \alpha \in Comp^- \subseteq A, \alpha \in rlvt_F(B)$, i.e., $Comp^- \subseteq rlvt_F(B)$.

Since $Comp^-$ is contained in $rlvt_F(B)$, $(Comp \cup Comp^-, R_{Comp} \cup I_{Comp})$ is a conditioned sub-framework with respect to $(rlvt_F(B), R_{rlvt_F(B)})$. For all $E_1 \in (rlvt_F(B), R_{rlvt_F(B)})$, we get a partially assigned sub-framework: $(Comp \cup Comp^-, R_{Comp} \cup I_{Comp})^{E_1}$. After the extensions of all partially assigned sub-frameworks are obtained, we combine them with the extensions of $(rlvt_F(B), R_{rlvt_F(B)})$ to form the extensions of $(rlvt_F(B'), R_{rlvt_F(B')})$. Formally, for all $\sigma \in \{adm, co, pr, gr\}$, we have

$$
\begin{aligned}
\mathcal{E}_\sigma((rlvt_F(B'), R_{rlvt_F(B')})) = \{E_1 \cup E_2 \mid \\
(E_1 \in \mathcal{E}_\sigma((rlvt_F(B), R_{rlvt_F(B)}))) \wedge \\
(E_2 \in \mathcal{E}_\sigma((Comp \cup Comp^-, R_{Comp} \cup I_{Comp}))^{E_1})\}
\end{aligned}
\tag{8.1}
$$

Formula 8.1 indicates that the partial semantics of F with respect to B' is obtained by extending the partial semantics of F with respect to B.

Example 8.6. Continue Example 8.5. The conditioned sub-framework for $Comp_1$ is represented as $(Comp_1 \cup Comp_1^-, R_{Comp_1} \cup I_{Comp_1})$.

Under preferred semantics, $\mathcal{E}_{pr}((rlvt_{F_{8.1}}(B_4), R_{rlvt_{F_{8.1}}(B_4)})) = \{E_{4,1}, E_{4,2}\}$, in which $E_{4,1} = \{a_1, a_3\}$, and $E_{4,2} = \{a_2\}$.

According to every extension of $(rlvt_{F_{8.1}}(B_4), R_{rlvt_{F_{8.1}}(B_4)})$, each argument in $Comp_1^-$ is assigned with a unique status, and we get $Comp_1^-[E_{4,1}]$ and $Comp_1^-[E_{4,2}]$, in which $Comp_1^-[E_{4,1}] = (\{a_3\}, \emptyset, \emptyset)$ and $Comp_1^-[E_{4,2}] = (\emptyset, \{a_3\}, \emptyset)$. The corresponding two partially assigned sub-frameworks (Figure 8.3) are represented as follows:

$$
(Comp_1 \cup Comp_1^-, R_{Comp_1} \cup I_{Comp_1})^{E_{4,1}}
$$
$$
(Comp_1 \cup Comp_1^-, R_{Comp_1} \cup I_{Comp_1})^{E_{4,2}}
$$

Under preferred semantics, the extensions of the two partially assigned sub-frameworks are computed:

$$
\mathcal{E}_{pr}((Comp_1 \cup Comp_1^-, R_{Comp_1} \cup I_{Comp_1})^{E_{4,1}}) = \{E_{4,3}\}, \text{ in which}
$$
$$
E_{4,3} = \{a_{12}\};
$$

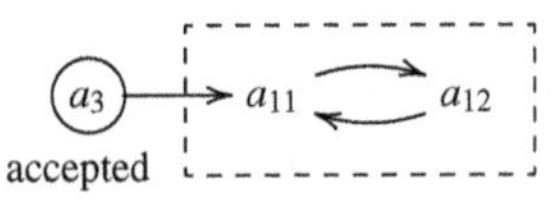

$$(1) \; (Comp_1 \cup Comp_1^-, R_{Comp_1} \cup I_{Comp_1})^{E_{4,1}}$$

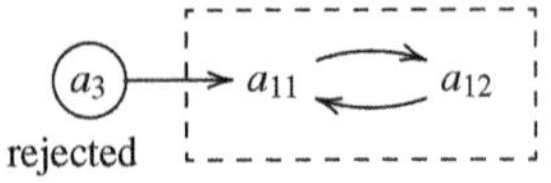

$$(2) \; (Comp_1 \cup Comp_1^-, R_{Comp_1} \cup I_{Comp_1})^{E_{4,2}}$$

Figure 8.3 The status of a_3 is assigned with "accepted" (according to $E_{4,1}$) in the first PASF, and with "rejected" (according to $E_{4,2}$) in the second PASF.

$$\mathscr{E}_{pr}((Comp_1 \cup Comp_1^-, R_{Comp_1} \cup I_{Comp_1})^{E_{4,1}}) = \{E_{4,4}, E_{4,5}\}, \text{ in which}$$
$$E_{4,4} = \{a_{11}\},$$
$$E_{4,5} = \{a_{12}\}.$$

According to Formula 8.1, the partial semantics of $F_{8.1}$ with respect to B_5 is as follows: $\mathscr{E}_{pr}((rlvt_{F_{8.1}}(B_5), R_{rlvt_{F_{8.1}}(B_5)})) = \{E_{5,1}, E_{5,2}, E_{5,3}\}$, where $E_{5,1} = E_{4,1} \cup E_{4,3} = \{a_1, a_3, a_{12}\}$, $E_{5,2} = E_{4,2} \cup E_{4,4} = \{a_2, a_{11}\}$, $E_{5,3} = E_{4,2} \cup E_{4,5} = \{a_2, a_{12}\}$.

8.3.3 Combinability of Partial Semantics

Given $F = (A, R)$ and $B, C \subseteq A$, the *combinability* of partial semantics of argumentation can be informally expressed as follows:

> The combination of partial semantics of F with respect to B and the one with respect to C is equal to the partial semantics of F with respect to $B \cup C$.
> Formally, we have the following definition.

Definition 8.3. Given $F = (A, R)$, $B, C \subseteq A$, let $Int = rlvt_F(B) \cap rlvt_F(C)$ denote the *intersection* of $rlvt_F(B)$ and $rlvt_F(C)$. Under a semantics $\sigma \in \{adm, co, pr, gr\}$, the combination of partial semantics of F with respect to B and the one with respect to C, denoted as $CombExt_\sigma((rlvt_F(B \cup C), R_{rlvt_F(B \cup C)}))$, is defined as follows:

$$CombExt_\sigma((rlvt_F(B \cup C), R_{rlvt_F(B \cup C)})) =_{def} \{E_1 \cup E_2 \mid$$
$$(E_1 \in \mathscr{E}_\sigma((rlvt_F(B), R_{rlvt_F(B)}))) \wedge$$
$$(E_2 \in \mathscr{E}_\sigma((rlvt_F(C), R_{rlvt_F(C)}))) \wedge$$
$$(E_1 \cap Int = E_2 \cap Int)\}$$

According to Proposition 5.2, under a semantics $\sigma \in \{adm, co, pr, gr\}$, it holds that $\mathcal{E}_\sigma((rlvt_F(B \cup C), R_{rlvt_F(B \cup C)})) = CombExt_\sigma((rlvt_F(B \cup C), R_{rlvt_F(B \cup C)}))$. According to this property, after we get the partial semantics of F with respect to some subsets of A respectively, we may get the partial semantics of a larger subset of arguments by means of semantics combination.

Example 8.7. Given $B_1 = \{a_9, a_{10}\}$, $B_5 = \{a_{11}, a_{12}\}$, according to Examples 8.4 and 8.6, we have:

$$\mathcal{E}_{pr}((rlvt_{F_{8.1}}(B_1), R_{rlvt_{F_{8.1}}(B_1)})) = \{E_{1,1}, E_{1,2}, E_{1,3}\}, \text{ in which}$$
$$E_{1,1} = \{a_1, a_9\},$$
$$E_{1,2} = \{a_1, a_{10}\},$$
$$E_{1,3} = \{a_2, a_{10}\},$$
$$\mathcal{E}_{pr}((rlvt_{F_{8.1}}(B_5), R_{rlvt_{F_{8.1}}(B_5)})) = \{E_{5,1}, E_{5,2}, E_{5,3}\}, \text{ in which}$$
$$E_{5,1} = \{a_1, a_3, a_{12}\},$$
$$E_{5,2} = \{a_2, a_{11}\},$$
$$E_{5,3} = \{a_2, a_{12}\}.$$

Let $Int_1 = rlvt_{F_{8.1}}(B_1) \cap rlvt_{F_{8.1}}(B_5)$. $Int_1 = \{a_1, a_2\}$. Under preferred semantics, the partial semantics of $F_{8.1}$ with respect to $B_1 \cup B_5$ can be obtained by means of semantics combination:

$$\mathcal{E}_{pr}((rlvt_{F_{8.1}}(B_1 \cup B_5), R_{rlvt_{F_{8.1}}(B_1 \cup B_5)})) = \{E_{6,1}, E_{6,2}, E_{6,3}, E_{6,4}\}, \text{ in which}$$
$$E_{6,1} = = E_{1,1} \cup E_{5,1} = \{a_1, a_3, a_9, a_{12}\},$$
$$E_{6,2} = E_{1,2} \cup E_{5,1} = \{a_1, a_3, a_{10}, a_{12}\},$$
$$E_{6,3} = E_{1,3} \cup E_{5,2} = \{a_2, a_{10}, a_{11}\},$$
$$E_{6,4} = E_{1,3} \cup E_{5,3} = \{a_2, a_{10}, a_{12}\}.$$

8.4 Empirical Investigation

In this section, we conduct an empirical investigation on the properties of using ASP to compute the partial semantics of argumentation.

As presented in Chapter 3, different encodings for various argumentation semantics have been proposed, including [3–5]. In this chapter, we take Egly et al's encodings [4] as an example, and consider the encoding under preferred semantics.

In order to evaluate the properties of computing the partial semantics of argumentation, we developed a Java program based on a Java wrapper for DLV.[1] The DLV Wrapper is an

---

[1] http://www.dlvsystem.com/dlvsystem/index.php/DLV_WRAPPER.

Object-Oriented library that "wraps" up the DLV system[2] in a Java program. In other words, the DLV Wrapper acts as an interface between Java programs and the DLV system.

The program first generates at random an argumentation framework (a defeat graph) with a given edge density (1%, 2%, ..., 20%) and a given size of the graph (50, 55, ..., 100). Here, the *edge density* of a defeat graph $F = (A, R)$ can be defined as follows:

$$density = \frac{|R|}{|A| * (|A| - 1)} * 100\% \tag{8.2}$$

Informally, the edge density of a given defeat graph denotes the percentage of its edges out of all possible edges between nodes.

Then, given the number of arguments to be queried (1, 10 percent of the number of nodes, or 20 percent of the number of nodes), the set of arguments to be queried is generated at random. And then, according to the set of arguments to be queried, the set of arguments belonging to the unattacked set is generated, and the sub-framework induced by the unattacked set is constructed subsequently. Finally, the preferred extensions of the argumentation framework, and of the sub-framework induced by the unattacked set, are computed respectively.

This program was tested on a machine with an Intel CPU running at 1.86 GHz and 1.98 GB RAM. The following three aspects of execution time were measured: the time for generating an unattacked set and constructing a sub-framework induced by the unattacked set, the time for computing the extensions of the argumentation framework, and the time for computing the extensions of the sub-framework induced by the unattacked set.

For convenience, the following notions are used: "di%-whole [n]", denoting the time for computing the extensions of the (whole) argumentation framework whose edge density is i% and the number of nodes is n; "di%-part-q1 [n]" (respectively "di%-part-q10% [n]", and "di%-part-q20% [n]"), denoting the time for generating an unattacked set, constructing a sub-framework induced by the unattacked set, and computing the extensions of the sub-framework, of the argumentation framework whose edge density is i% and the number of nodes is n, when the number of arguments to be queried is 1 (respectively, 10 percent of the number of nodes, and 20 percent of the number of nodes).

For each configuration with a certain edge density and node number, the program was executed 20 times, and the average execution time of different aspects was obtained.

Table 8.1 shows the execution time in the case where the edge density of argumentation frameworks is 2% and the number of nodes is 85. We found that in some cases, the execution time might be much higher than the ones in other cases (in Table 8.1, the execution time of

---

[2] DLV is an efficient ASP solver. Please refer to the following web page for details: http://www.dbai.tuwien.ac.at/proj/dlv/.

Table 8.1 The execution time of the case where the edge density of argumentation frameworks is 2% and the number of nodes is 85.

No.	d2%-whole [85] (seconds)	d2%-part-q1 [85] (seconds)	d2%-part-q10% [85] (seconds)	d2%-part-q20% [85] (seconds)
1	11.625	1.297	1.547	5.735
2	42.188	5.984	8.578	9.344
3	15.812	0.234	3.469	3.875
4	6.469	1.078	2.250	1.765
5	**119.375**	**52.334**	**71.765**	**93.969**
6	31.266	0.437	4.625	7.656
7	65.906	0.235	19.468	16.391
8	**689.672**	**351.187**	**370.516**	**371.297**
9	26.328	4.547	3.968	9.453
10	13.906	2.844	3.484	6.141
11	21.140	0.219	8.016	9.140
12	58.110	15.250	15.203	18.109
13	21.937	0.235	9.203	16.281
14	39.328	22.203	27.860	34.562
15	10.797	0.219	6.781	8.734
16	72.234	9.735	16.515	17.125
17	56.172	0.234	16.844	22.484
18	28.188	4.281	4.750	4.891
19	22.250	1.157	2.375	2.703
20	22.313	8.344	8.687	10.094
Avg.	68.750	24.130	30.295	33.487
Rev. Avg.	**31.441**	**4.362**	**9.089**	**11.359**

rows 5 and 8 is much higher than that of others). In order to reflect the situations in most cases, the average execution time is revised by dropping the cases where the execution time of computing the extensions of the (whole) argumentation framework is 3 times more than the average of remaining cases.

In Figure 8.4, we present the execution time when the number of nodes is from 50 to 100 and the edge density of argumentation frameworks is 1%, 2%, 3% and 20%, respectively. When the edge density of an argumentation framework is 1%, which means that when the size of the argumentation framework is 100, the ratio of nodes to edges is 1:1, the average time for computing the partial semantics is much lower than that of computing the semantics of the whole argumentation framework. With the increase of edge density, the average time for computing the partial semantics becomes closer and closer to that of computing the semantics of the whole argumentation framework. When the number of nodes is from 50 to 100 and the edge density is no less than 6%, the average time for computing the partial semantics is almost equal to that of computing the semantics of the whole argumentation framework. Then, with the increase of edge density, the relation between these two kinds of execution time remains the same. This phenomenon can be well explained by Figure 8.5 and Table 8.2.

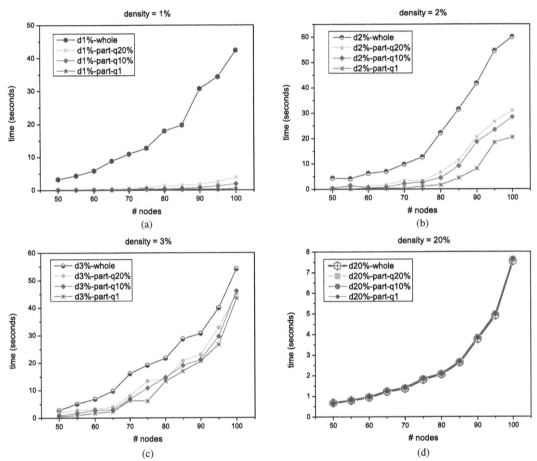

Figure 8.4 Plots showing the average execution time (revised by dropping some cases where the execution time of computing the extensions of the whole argumentation framework is 3 times more than the average execution time in remaining cases) when the edge density of argumentation frameworks is 1%, 2%, 3% and 20%, respectively.

Figure 8.5(a) and (b) shows that in the cases where the edge density of an argumentation framework is no less than 6%, the size of the unattacked set is equal to the number of all nodes in the argumentation framework. In these cases, it is obvious that the execution time of computing the partial semantics is no less than that to compute the semantics of the whole argumentation framework. Meanwhile, as shown in Table 8.2, since in various cases, the execution time for generating an unattacked set and constructing a sub-framework induced by the unattacked set is very low, it can be neglected. As a result, in Figure 8.4(d), when the edge density is 20%, the four kinds of execution time overlap.

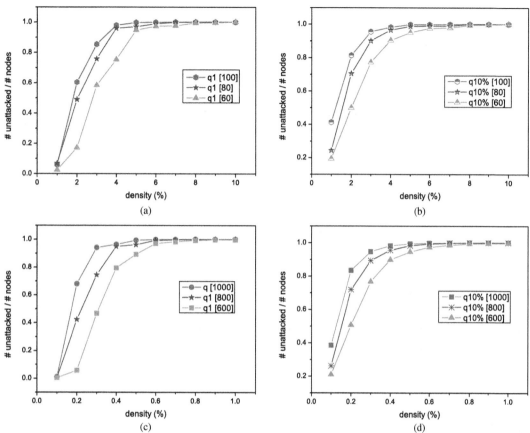

Figure 8.5 Plots showing the relation between the ratio of the average size of the unattacked set to the number of the nodes, of an argumentation framework, and the edge density of the argumentation framework. The notation "q1 [*n*]" ("q10% [*n*]") denotes the case where the number of nodes in the argumentation framework is *n*, and the number of arguments to be queried is 1 (respectively, 10 percent of the number of nodes).

In addition, it is interesting to note that when the number of nodes is from 500 to 1000 and the edge density of an argumentation framework is no less than 0.6%, the size of the unattacked set is nearly equal to the number of all nodes in the argumentation framework (as shown in Figure 8.5(c) and (d)). The reason behind this phenomenon is the fact that when the number of nodes is 1000 and the edge density is 0.6%, the ratio of the number of edges to the number of nodes is 6:1, which is equal to the one in the case where the number of nodes is 100 and the edge density is 6%.

According to the above analysis, we may conclude that the computation of partial semantics of argumentation has the following two properties:

Table 8.2 The execution time for generating an unattacked set and constructing a sub-framework induced by the unattacked set. The notation "d*i*%-part$_s$-q10%" denotes the time for generating an unattacked set and constructing a sub-framework induced by the unattacked set, of the argumentation framework whose edge density is *i*%, when the number of arguments to be queried is 10 percent of the number of nodes.

# nodes	d1%-part$_s$-q10% (seconds)	d2%-part$_s$-q10% (seconds)	d3%-part$_s$-q10% (seconds)	d5%-part$_s$-q10% (seconds)	d10%-part$_s$-q10% (seconds)
50	0.003	0.001	0.009	0.008	0.015
55	0.001	0.005	0.006	0.009	0.014
60	0.002	0.002	0.007	0.010	0.018
65	0.001	0.004	0.007	0.011	0.018
70	0.002	0.003	0.010	0.012	0.024
75	0.003	0.003	0.007	0.014	0.026
80	0.003	0.004	0.009	0.014	0.028
85	0.003	0.011	0.011	0.016	0.031
90	0.003	0.007	0.012	0.017	0.033
95	0.003	0.011	0.014	0.021	0.037
100	0.005	0.011	0.016	0.025	0.041

(i) The method of computing the partial semantics of argumentation is efficient when the defeat graphs of argumentation frameworks are sparse, especially when the ratio of the number of edges to the number of nodes is less than 2:1.

(ii) When the ratio of the number of edges to the number of nodes is more than 6:1, the average time for computing the partial semantics is almost equal to that of computing the semantics of the whole argumentation framework.

8.5 Conclusions

In this chapter, based on the idea of local computation, we have presented the definition and three basic properties of the partial semantics of argumentation, and conducted an empirical investigation on the properties of computing the partial semantics of argumentation. Since in many situations, only the status of some part of arguments in an argumentation framework is necessary to be figured out, when the defeat graphs of argumentation frameworks are sparse (more specifically, when the ratio of the number of edges to the number of nodes, of a defeat graph, is less than 2:1), the partial computation of argumentation semantics is apparently more efficient.

References

[1] S. Modgil, M. Caminada, Proof theories and algorithms for abstract argumentation frameworks, in: Argumentation in Artificial Intelligence, Springer, 2009, pp. 105–129.

[2] G. Boella, D.M. Gabbay, A. Perotti, L. van der Torre, S. Villata, Conditional labelling for abstract argumentation, in: Proceedings of the First International Workshop on the Theory and Applications of Formal Argumentation, 2012, pp. 232–248.

[3] T. Wakaki, K. Nitta, Computing argumentation semantics in answer set programming, in: Proceedings of the 22nd Annual Conference of the Japanese Society for Artificial Intelligence, 2008, pp. 254–269.

[4] U. Egly, S.A. Gaggl, S. Woltran, Answer-set programming encondings for argumentation frameworks, Argument and Computation 1 (2) (2010) 147–177.

[5] J.C. Nieves, U. Cortés, M. Osorio, Preferred extensions as stable models, Theory and Practice of Logic Programming 8 (4) (2008) 527–543.

CHAPTER 9

Conclusions and Future Work

Chapter Outline

9.1 Conclusion

This book has presented fundamental theories for efficient computation of argumentation semantics. Under admissible, complete, preferred and grounded semantics, we have proved that the local semantics could be combined to form the global semantics, of an argumentation framework. The idea of local computation and global computation is somewhat similar to that of the *divide and conquer* paradigm. A divide and conquer algorithm works by recursively breaking down a problem into two or more sub-problems of the same (or related) type, until these become simple enough to be solved directly. The solutions to the sub-problems are then combined to give a solution to the original problem.

Based on the idea of local computation, three approaches for efficient computation of argumentation semantics from three different perspectives (static semantics of argumentation, dynamic semantics of argumentation and partial semantics of argumentation) have been formulated. The efficiency of these approaches is achieved by exploiting the following important properties of computation. The first property is that in some argumentation frameworks that do not belong to any tractable classes, some of their sub-frameworks might be tractable, while others might not, but have smaller sizes. This property is called the *local tractability* of argumentation frameworks. The second property is that when an argumentation framework is updated, only the status of some part of arguments is affected. This property is called the *local dynamics* of argumentation frameworks. By exploiting this property, some results of previous computation could be reused. The third property is that when querying the status of some specific arguments, only a part of arguments is relevant.

Efficient Computation of Argumentation Semantics. http://dx.doi.org/10.1016/B978-0-12-410406-8.00009-9

9.2 Future Work

Based on the theories and approaches presented in this book, the following directions are worth being investigated.

First, developing more general relations between local semantics and global semantics. In this book, we have only studied the relations under admissible, complete, preferred and grounded semantics. When considering some other semantics, such as ideal, eager, stable and semi-stable, etc., an important problem is that the mappings between local semantics and global semantics might be unsound and/or incomplete. Hence, further research is necessary to handle this problem. Meanwhile, as mentioned in Chapter 5, since there are two kinds of sub-frameworks (conditioned and unconditioned) and various dependence relations between different sub-frameworks, there are many types of combinations of sub-frameworks. In this book, we have only formulated three of them, while some other types of combinations are unexplored.

Second, extending the relations between local semantics and global semantics to multiple argumentation frameworks. In this book, we have considered the local semantics and the semantics combination within an argumentation framework. In multi-agent scenarios, each agent may have an argumentation framework, which may interact with other argumentation frameworks. In [1], given a set of distinct argumentation frameworks from different agents, they are expanded respectively into partial systems over the set of all arguments considered by the group of agents. Then, a merging operator is used to produce a set of argumentation systems that are as close as possible to the partial systems (to realize a kind of consensus). And then, the acceptability of a set of arguments at the group level is obtained by selecting the extensions of a set of produced (merged) argumentation frameworks at the local level. Currently, how to relate the dynamics of local semantics to the dynamics of global semantics in multi-agent scenarios is still an open problem.

Third, integrating the efficient approaches presented in the book to some other efficient algorithms. Different from the approaches of local computation, there are some specific algorithms for improving the efficiency of computing the status of arguments. For instance, in [2], Nofal et al. presented a case study on experimental algorithms in the context of an instance of extended argumentation frameworks, and analysed the efficiency of three different algorithms for deciding the acceptability of an argument with respect to a set of arguments; in [3], they proposed a more efficient algorithm for enumerating all preferred extensions, by utilizing further labels to improve labels' transitions. These algorithms all treat argumentation frameworks as monolithic entities. It is interesting to integrate these kinds of algorithms to our local computation approaches.

Fourth, studying the worse-case (algorithm independent) computational complexity of the local computation-based approaches. In existing literature, the computational complexity of argumentation frameworks satisfying graph-theoretic constraints has been studied [4]. However, similar to the existing algorithms, all argumentation frameworks are treated as monolithic entities. So, from the perspective of local computation, it is interesting to explore how the computational complexity of an argumentation framework is related to some of its local properties, such as the size of maximal SCC, the number of affected arguments and the number of relevant arguments, etc.

Fifth, applying the local computation-based approaches to some extended argumentation frameworks. Besides the Dung's abstract argumentation framework, there are some other argumentation frameworks extended from different perspectives, such as bipolar argumentation frameworks [5], weighted argumentation systems [6], and value-based argumentation frameworks [7], etc. The approaches and algorithms for computing the extensions of such argumentation frameworks have not been extensively studied. The idea of local computation might be applicable with respect to this new direction.

Sixth, developing local computation-based approaches for instantiated argumentation systems. In this book, we have only considered the computation of Dung's abstract argumentation framework. For some instantiated argumentation systems, such as ASPIC$^+$ [8] and first-order logic-based argumentation systems [9], it seems that the local computation is related not only to the attack relation, but also to the support relation, between arguments. Further investigation in this direction could be challenging and interesting.

References

[1] S. Coste-Marquis, C. Devred, S. Konieczny, M. Lagasquie-Schiex, P. Marquis, On the merging of Dung's argumentation systems, Artificial Intelligence 171 (10–15) (2007) 730–753.

[2] S. Nofal, P. Dunne, K. Atkinson, Towards experimental algorithms for abstract argumentation, in: Proceedings of the Fourth International Conference on Computational Models of Argument, 2012, pp. 217–228.

[3] S. Nofal, P. Dunne, K. Atkinson, On preferred extension enumeration in abstract argumentation, in: Proceedings of the Fourth International Conference on Computational Models of Argument, 2012, pp. 205–206.

[4] P.E. Dunne, Computational properties of argument systems satisfying graph-theoretic constraints, Artificial Intelligence 171 (10–15) (2007) 701–729.

[5] C. Cayrol, M.-C. Lagasquie-Schiex, On the acceptability of arguments in bipolar argumentation frameworks, in: Proceedings of the Eighth European Conference on Symbolic and Quantitative Approaches to Reasoning with Uncertainty, 2005, pp. 378–389.

[6] P.E. Dunne, A. Hunter, P. McBurney, S. Parsons, M. Wooldridge, Weighted argument systems: basic definitions, algorithms, and complexity results, Artificial Intelligence 175 (10–15) (2011) 457–486.

[7] T.J.M. Bench-Capon, Persuasion in practical argument using value-based argumentation frameworks, Journal of Logic and Computation 13 (3) (2003) 429–448.

[8] H. Prakken, An abstract framework for argumentation with structured arguments, Argument & Computation 1 (2) (2010) 93–124.

[9] V. Efstathiou, A. Hunter, Algorithms for generating arguments and counter-arguments in propositional logic, International Journal of Approximate Reasoning 52 (6) (2011) 672–704.

Index

Printed and bound by CPI Group (UK) Ltd, Croydon, CR0 4YY

03/10/2024

01040326-0003